Introduction to
Queueing Networks

Introduction to Queueing Networks

E. Gelenbe

Université de Paris Sud, Paris, France

and

G. Pujolle

Université Pierre et Marie Curie, Paris, France

Translated by

J. C. C. Nelson

University of Leeds, Leeds, UK

JOHN WILEY & SONS

Chichester · New York · Brisbane · Toronto · Singapore

This work is the translation of the French book
by Gelenbe and Pujolle:
Introduction aux réseaux de files d'attente
published by Editions Eyrolles, Paris
Copyright © 1987 by John Wiley & Sons Ltd.

Library of Congress Cataloging-in-Publication Data:

Gelenbe, E., 1945–
    Introduction to queueing networks.

    Translation of: Introduction aux réseaux de files
d'attente.
    Includes bibliographies and index.
    1. Queueing theory.   I. Pujolle, G., 1949–
II. Title.
T57.9.G4513   1987   519.8′2   86–13197

ISBN 0 471 90464 3

British Library Cataloguing in Publication Data:

Gelenbe, E.
    Introduction to queueing networks.
    1. Computer networks   2. Queueing
theory
I. Title   II. Pujolle, G.   III. Introduction
aux résaux de files d'attente. *English*
004.6   TK5105.5

ISBN   0 471 90464 3

Phototypesetting by Thomson Press (India) Limited, New Delhi
Printed and bound in Great Britain

Contents

v

INTRODUCTION

Queues and queueing systems have been the subject of considerable research since the appearance of the first telephone systems. In addition to models originating in biology and genetics (branching processes), they have been the principal examples of realistic discrete state space random processes.

In the years immediately following the Second World War, the problems of operational research, that is inventory and production control, aroused a new interest in this subject area. It was rapidly discovered that models of the reliability of complex systems could well be formulated in terms of queues (arrivals of breakdowns and repair sevices). Moreover, these two aspects have given rise to an abundant literature on the optimization problems for particular queueing models.

The modelling of computer systems and data transmission systems opened the way, in the sixties, to studies of queues characterized by complex service disciplines and have created the need to analyse interconnected systems. Progress in this area has been rapid and industrial applications have been widely accepted since the seventies. At present in the computer industry, queuing network models have resulted in software packages for the automatic solution of problems arising in the design of new computers and in the evaluation and improvement of existing systems.

The methods of queueing networks have always been a basic component of the study of communication systems. The widespread introduction of computers into these systems has introduced the use, in a systematic manner, of new results on queueing networks in studies of the performance of large communication networks.

There has also been a renewal of interest by mathematicians in a subject which has developed outside the traditional mainstream of probability theory. The combination of fundamental considerations and practical problems is the best guarantee of the vitality of the subject.

This book makes no pretension to be exhaustive in an area which includes a large number of significant results and contributions. In the choice of subjects which are presented it has been necessary to make a selection which has caused us to exclude interesting theoretical aspects (such as a detailed study of a queue with one server, which is well treated elsewhere [6]).

The presentation of this book is biased towards the study of queueing networks since they lead, in our opinion, to the most interesting possibilities for application to data processing and communication systems. It is also necessary to mention the limited number of works on the subject: in the list of thirty basic books which we cite, there are only four or five which treat the subject and in a manner which is often superficial. With regard to the application examples presented in this work, we have very often chosen them in the area of data transmission networks. There, also, we have taken into consideration the fact that books which treat the performance evaluation of computer systems are more numerous.

Concerning the presentation of this book, it should be mentioned that Chapter 1 has been written with the object of giving the reader an outline of the methods used in the following chapters. We have developed, in a tutorial manner, the complementarity between the *deterministic* and *probabilistic* approaches; a large place has been reserved for the use, within the simple context of a queue with a single server, of methods arising in the theory of regenerative processes developed in [5]. A complete reading of this chapter, with the possible exception of Sections 1.6 and 1.7, is recommended even to readers interested primarily in applications.

Chapter 2 is devoted to the simplest queueing networks. A deterministic approach allows the introduction of the subject in an elementary but rigorous manner, and a substantial part is devoted to numerical algorithms which permit practical application. The probabilistic aspect is then introduced and the formal equivalence (that is to say in the sense of the equations obtained) with the deterministic case is shown.

Chapter 3 plays a less important role; here different particular cases (such as systems having a limited storage capacity) are treated to illustrate the theory. Nevertheless, we introduce the use of diffusion processes to the analysis of queueing systems.

Chapter 4 is concerned with virtually the most general queueing networks which it is known can be treated in an exact manner with solutions in 'product form'. This form of solution is very important since it permits decomposition of the joint probabilities of the states of the model into products of marginal probabilities.

This leads naturally to the examination of several different methods of solution which are presented in Chapter 5. It concerns, in particular, diffusion processes, the methods of decomposition and isolation, mean value analysis and also numerous examples of application examples.

The first five chapters are devoted almost entirely to a study of the *state* of queueing network as described by the position of customers at the service stations. A complementary approach, adopted in Chapter 6, concerns an examination of *flow* of customers through the network. This very recent and less well known approach is of a more theoretical nature and paves the way for new research directions.

Each chapter is accompanied by a bibliography containing books and articles of reference. In an appendix at the end of the book, numerous formulae suitable for practical use are proposed.

In the present chapter, we give two biographical lists, the first containing basic books, while the second deals with more specialized books or sources of additional information.

For the first list, references [1, 2, 8, 14, 17, 23, 26] are introductory works to the subject of which [1] is the easiest. The simple queue is treated in depth [6] and equally in [10, 13, 19, 21, 29, 30]. Relations with problems of operational research and reliability are treated in [15, 25, 27, 28] and relations with general methods of random processes in [3, 4, 7, 18] (see also [5]). Methods adapted to the study of computer systems (and in particular the method of decomposition) are presented in [9, 11, 12, 20, 22] and form a large part of the applications concerning data transmission networks. Results concerning systems with priorities appear in [16, 22].

Problems of evaluating the properties of computer systems by queueing networks are described in [34, 35, 39]. Models related to telephone systems appear in [32, 37, 40]. Aspects relative to the theory of point processes appear in [33] and other, more technical, aspects are treated for instance in [31, 36, 38, 41].

BIBLIOGRAPHY

1. Allen, A. O. (1978). *Probability, Statistics and Queueing Theory with Computer Science Applications,* Washington D. C.: Academic Press.
2. Beckman, P. (1968). *Introduction to Elementary Queueing Theory,* Boulder, Colo.: Golem Press.
3. Benes, V. E. (1963). *General Stochastic Processes in the Theory of Queues,* Mass: Addison, Wesley.
4. Borovkov, A. A. (1976). *Stochastic Processes in Queueing Theory,* New York: Springer-Verlag.
5. Cinlar, E. (1975). *Introduction to Stochastic Processes,* Englewood Cliffs, NJ: Prentice-Hall.
6. Cohen, J. W. (1969). *The Single Queue,* New York: American Elsevier.
7. Cohen, J. W. (1976). *On Regenerative Processes in Queueing Theory,* New York: Springer-Verlag.
8. Cooper, R. B. (1972). *Introduction to Queueing Theory,* New York: Macmillan.
9. Courtois, P. J. (1977). *Decomposability Queueing and Computer System Applications,* New York: Academic Press.
10. Cox, D. R., and Smith, W. L. (1961). *Queues,* London: Methuen.
11. Gelenbe, E., Labetoulle, J., Marie, R., Metivier, M., Pujolle, G., and Stewart, W. (1980). *Réseaux de files d'attente,* Paris: Éditions Hommes et Techniques.
12. Gelenbe E., and Mitrani I. (1980). *Analysis and Synthesis of Computer Systems,* London: Academic Press.
13. Genedenko, B. V., and Kovalenko, I. N. (1968). *Introduction to Queueing Theory,* Israel, Program for Scientific Translations, Jerusalem, Israel.
14. Gross D., and Harris C. M. (1974). *Fundamentals of Queueing Theory,* New York: Wiley.
15. Hillier, F. S., and Lieberman, G. J. (1967). *Introduction to Operations Research,* San Francisco: Holden-Day.
16. Jaiswal, N. (1968). *Priority Queues,* New York: Academic Press.
17. Kaufmann, A. (1972). *Méthodes et modèles de la recherche opérationnelle tome 1,* Paris: Dunod.
18. Kelly, F. P. (1979). *Reversibility and Stochastic Networks,* John Wiley.
19. Khintchine, A. Y. (1960). *Mathematical Methods in the Theory of Queueing,* London: Griffin.
20. Kleinrock, L. (1964). *Communication Nets,* McGraw-Hill, New York.
21. Kleinrock, L. (1975). *Queueing Systems, – Volume I: Theory,* John Wiley.
22. Kleinrock, L. (1976). *Queueing Systems, – Volume II: Computer Applications,* John Wiley.
23. Kobayashi, H. (1978). *Modeling and Analysis,* Addison Wesley.
24. Le Gall, P. (1962). *Les systèmes avec ou sans attente et les processus stochastiques, Tome I, –* Paris, France: Dunod.

25. Morse, P. M. (1958). *Queues, Inventories and Maintenance*, John Wiley.
26. Newell, G. F. (1972). *Applications of Queueing Theory*, London: Chapman and Hall.
27. Prabhu, N. U. (1965). *Queues and Inventories*, New York: Wiley.
28. Riordan, J. (1962). *Stochastic Service Systems*, New York: Wiley.
29. Saaty, T. L. (1961). *Elements of Queueing Theory with Applications*, New York: McGraw-Hill.
30. Takacs, L. (1962). *Introduction to the Theory of Queues*, Oxford University Press.
31. Bagchi, T. P., and Templeton, J. G. C. (1972). *Numerical Methods in Markov Chains and Bulk Queues*, Berlin: Springer-Verlag.
32. Benes, V. E. (1965). *Teletraffic Queueing Problem*, Mass.: Addison, Wesley.
33. Bremaud, P. (1981). *Dynamical Point Processes and Ito Systems in Communications and Queueing*, Berlin: Springer-Verlag.
34. Descloux, A. (1962). *Delay Tables for Finite- and Infinite-Source Systems*, New York: McGraw-Hill.
35. Ferrari, D. (1978). *Computer Systems Performance Evaluation*, Prentice-Hall, Engelwood Cliffs, N.J.
36. Ghosal, A. (1970). *Some Aspects of Queueing and Storage Systems*, Berlin: Springer-Verlag.
37. Haight, F. (1963). *Mathematical Theory of Traffic Flow*, New York: Academic Press.
38. Kaufmann, A. (1972). *Méthodes et modèles de la recherche opérationnelle – Tome I*, Paris: Dunod.
39. Svobodova, L. (1976). *Computer Performance Measurement and Evaluation Methods: Analysis and Applications*, New York: Elservier.
40. Syski, R. (1960). *Introduction to Congestion Theory in Telephone System*, London: Olivier and Boyd.
41. Teghem, J., Loris-Teghem J., and Lambotte, J. P. (1969). *Modèles d'attente M/M/1 et G1/M1 à arrivées et services en groups*. Berlin: Springer-Verlag.

CHAPTER 1

Queues with a Single Server

1.1 - INTRODUCTION

Most problems involving modelling of computer systems or data transmission networks deal with systems having multiple resources (central processing units, channels, memories, communication circuits, etc.) to be taken into account. This complex structure leads to the study of queueing networks, rather than simple queues with a single server.

Nevertheless, the study of a single queue is interesting for several reasons. On the one hand, understanding queueing phenomena is easier in the context of the simplest model. On the other hand, the simple queue is a useful framework for the development of the mathematical tools used in queues. Finally, 'seen from the outside' the entire computer or transmission system can be regarded as a unique server with a queue having a complex service discipline.

It is this last point of view which will be adopted in Section 1.2 where we shall examine, in a *deterministic* manner, a queue with a single server without specifying the service discipline (or the order in which service is allocated to customers). This deterministic approach will be used again in Section 1.4 where the links between deterministic and probabilistic results will be established from the properties of regenerative random processes. In Section 1.2 a general formula is obtained which computes the proportion of time spent by the queue in a given state, with very weak assumptions. Section 1.3 is devoted to the elementary 'M/M/1' queue—exponential distributions, Poisson processes, the Chapman–Kolmogorov equation and its stationary solution; here results similar to those of the preceding section are found again.

Section 1.4 starts with a proof of Little's formula for the deterministic case, as well as the case of a regenerative process, which establishes a simple and very useful link between the mean rate of arrival, the mean response time and the mean number in the queue. Before introducing Kendall's now classic notation for describing the properties of a simple queue, simple but important characteristics such as the distribution of the length of the queue at arrival and departure

1

times and the probability that the queue is empty are obtained. These results depend very little on the precise properties of the arrival and service processes.

Sections 1.5 and 1.6 show the classic results concerning queues with Poisson arrivals and a general service distribution or, conversely, a general arrival distributions but exponential services. The computation methods used make use of the properties of Markov renewal processes. Finally, in Section 1.7, certain elementary results are established for a queue with a single server where the arrival and service distributions are general (a 'GI/GI/1' system).

1.2 - DETERMINISTIC APPROACH TO A QUEUE WITH A SINGLE SERVER

In this section, we are interested in a very simple system and a deterministic analysis of its behaviour. An analogy will also be established between certain properties of its deterministic behaviour and those of its random behaviour.

A series of 'customers' carrying successive numbers $1, 2, 3, \ldots$ etc., arrive at instants $a_1 < a_2 < a_3 < \cdots$ at the queue of Figure 1.1. The server deals with them in some order; the customers line up in the queue and it is always the one at the head who is being served.

Let $N(t)$ be the number of customers waiting in the queue plus the one who is being served at time t and consider a time interval $[a, b]$ such that $N(a) = N(b) = 0$ (see Figure 1.2).

For this time interval let $T(n)$ be the time spent by the queue in the state $N(t) = n$;

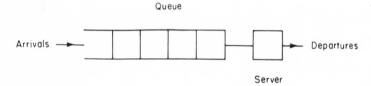

Queue

Arrivals ⟶

Departures

Server

Figure 1.1 Simple queue

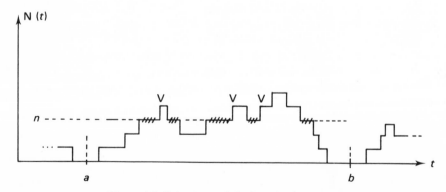

Figure 1.2 Behaviour of the length of a queue

that is the hatched regions in Figure 1.2. $\alpha(n)$ is the number of arrivals of customers when the length of the queue is n (indicated by V) and $\beta(n)$ is the number of departures when the state is n, still in the same interval $[a, b]$. Since $N(a) = N(b) = 0$, it is necessary that:

$$\alpha(n) = \beta(n + 1).$$

Let $T = b - a$ be the duration of this interval. Hence:

$$\frac{\alpha(n)}{T} = \frac{\beta(n + 1)}{T}, \quad n = 0, 1, 2, \ldots.$$

Let $p(n)$ be the proportion of time spent in state n, in the interval in question:

$$p(n) = \frac{T(n)}{T}.$$

Hence:

$$p(n)\frac{\alpha(n)}{T(n)} = p(n + 1)\frac{\beta(n + 1)}{(n + 1)}.$$

To simplify these formulae, let $\lambda(n)$ be the number of arrivals in unit time when the queue is of length n in the interval $[a, b]$:

$$\lambda(n) = \alpha(n)/T(n)$$

and $\mu(n)$ the number of departures in unit time:

$$\mu(n) = \beta(n)/T(n).$$

Hence we have the recurrence:

$$p(n + 1) = [\lambda(n)/\mu(n + 1)]p(n), \quad n = 0, 1, \ldots$$

of which the solution is:

$$p(n) = p(0) \prod_{i=1}^{n} \frac{\lambda(i - 1)}{\mu(i)}, \quad n = 1, 2, \ldots \tag{1.2.1}$$

Therefore the proportions of time spent in each state must satisfy (1.2.1) and this formula will be applicable to all time intervals which start and finish with an 'empty' state of the queue. $p(0)$ can easily be determined since we must obtain:

$$\sum_{n=0}^{\infty} \frac{T(n)}{T} = \sum_{n=0}^{\infty} p(n) = 1$$

which gives:

$$p(0) = \left[1 + \sum_{n=1}^{\infty} \prod_{i=1}^{n} (\lambda(i - 1)/\mu(i))\right]^{-1}. \tag{1.2.2}$$

It must be emphasized that the quantities $\lambda(n)$ and $\mu(n)$ are *measurable* experimentally; but formula (1.2.1) does not *predict* the queue's behaviour outside the precise interval which starts at time a and finishes at time b.

Nevertheless, property (1.2.1) occurs in certain probabilistic models which we shall examine subsequently and which are used to *predict* behaviours in known statistical conditions.

1.3 - THE EXPONENTIAL DISTRIBUTION AND QUEUES WITH A SINGLE SERVER

When it is desirable to *predict* the performance of a system outside an observation period during which all the data concerned are accessible, it becomes necessary to make certain assumptions concerning its behaviour. The probabilistic assumptions which we shall make in this section and in those which follow lead to predictions concerning the systems which we shall analyse. The link between these probabilistic assumptions and measurements on actual systems is established with the aid of statistics.

The results of Section 1.2 concern a particular interval $[a, b]$ and a deterministic behaviour. The results which we obtain in Section 1.3 concern an infinite period of time and a (probable) set of realizations.

In this section, we assume that the durations $I_1 = a_1 - 0$, $I_2 = a_2 - a_1, \ldots$, are random variables, that is they consist of quantities whose exact values are not known but for which a probability distribution can be defined. The variables $S_1, S_2, S_3, \ldots$ represent successive service times and are also assumed to be random.

Also, we assume that the time intervals between successive arrivals, or *interarrivals*, $I_1, I_2, I_3, \ldots$ are all distributed according to the exponential distribution:

$$P\{I_j < x\} = 1 - e^{-\lambda x}, \quad \text{for all } j \geqslant 1,$$

the same assumption is made for the service times:

$$P\{S_i < x\} = 1 - e^{-\mu x}, \quad i \geqslant 1,$$

where λ and μ (the parameters of the distributions) are real positive and finite. On the other hand we suppose that for $i \neq j$:

$$P\{I_i < x \quad \text{and} \quad I_j < y\} = P\{I_i < x\}P\{I_j < y\}$$

that is to say the interarrivals are *independent*. We also assume that the service times are independent of each other and that they are independent of the interarrivals.

1.3.1 'Memoryless' property of the exponential distribution

The exponential distribution has one particularly interesting property which is one of the factors explaining its popularity. Suppose that we are dealing with a time bomb which explodes automatically after a time X distributed according to an exponential distribution.

$$P\{X < x\} = 1 - e^{-\lambda x}, \quad \infty > \lambda > 0.$$

We trigger the mechanism at time $t = 0$ to cause an explosion at time $t = X$. At an intermediate time $t = y$, before the explosion occurs, we would like to know the time remaining before the explosion!

This simply means that we wish to know the distribution of $X - y$ knowing that $X > y$ since the explosion has not occurred at time $t = y$. We calculate:

$$P\{X - y < x | X > y\} = P\{y < X < y + x\}/P\{X > y\},$$
$$= \frac{1 - e^{-\lambda(y+x)} - (1 - e^{-\lambda y})}{e^{-\lambda y}},$$
$$= 1 - e^{-\lambda x} = P\{X < x\} \tag{1.3.1}$$

and we discover that the fact that the explosion has not occurred up to time t allows us to establish simply that $X - y$ has the same distribution as X. This is called the *Markovian* or '*memoryless*' property of the exponential distribution.

In fact it can be proved that if a continuous positive random variable has the property (1.3.1), then its distribution is exponential.

1.3.2 Analysis of a queue with exponential interarrivals and services

The queue being studied has been defined as a system in which the times between successive arrivals and the service times are mutually independent random variables distributed according to the exponential distribution. We approach its analysis in a manner analogous to that adopted in Section 1.2 and we let $N(t)$ be the instantaneous number of customers in the queue, including the one who is being served.

Consider the existence of distinct times a and b at which the queue is empty ($N(a) = N(b) = 0$), but chosen such that $N(t) > 0$ for $t = a^+$ and $t = b^+$: that is, at times a^+ and b^+, an arrival occurs. We require also that the queue becomes empty once between a and b. This leads to the arrangement presented in Figure 1.3.

Let $\pi_{i,j}$ be the probability of passing from state i to state j after an arrival at or departure from the queue.

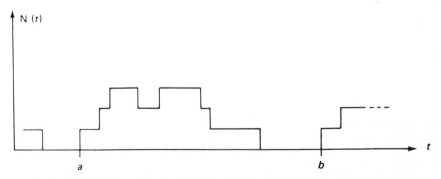

Figure 1.3 Example of the evolution of the number of customers in the queue

Let $m(n)$ be the mean number of times that the queue passes through state n between a and b. Since to reach state n, one must arrive either from state $n+1$ or from state $n-1$, one can immediately write for $n > 0$:

$$m(n) = \pi_{n+1,n} m(n+1) + \pi_{n-1,n} m(n-1), \tag{1.3.2}$$

a formula which will be verified if the passage from state i to state j depends only on i and j. This can be proved due to the 'memoryless' property of the exponential law of interarrivals and service durations.

Suppose that one is in a state $i > 0$ at a time t; the passage to state $i+1$ will occur at a time $t + y$ and between t and $t + y$ there will be no departures. The passage from i to $i - 1$ will be made in the same way.

So due to the memoryless property:

$$\pi_{i,i+1} = \int_{y=0}^{\infty} P\{y < I < y + dy\} P\{S > y\},$$

$$= \int_{0}^{\infty} dy \, \lambda e^{-\lambda y} e^{-\mu y} = \frac{\lambda}{\lambda + \mu},$$

where I and S denote the classes of interarrivals and service times. Also:

$$\pi_{i,i-1} = \int_{y=0}^{\infty} P\{I > y\} P\{y < S < y + dy\},$$

$$= \frac{\mu}{\lambda + \mu}.$$

One obtains for $n > 0$;

$$m(n) = \frac{\mu}{\lambda + \mu} m(n+1) + \frac{\lambda}{\lambda + \mu} m(n-1). \tag{1.3.3}$$

Let us now examine $T(n)$ which, in spite of the notation of Section 1.2, will be the *mean* time spent in state n between a and b. $T(n)$ is the mean time spent in state n on *each* stay in that state multiplied by $m(n)$. But thanks to the memoryless property, the time spent in a state $n > 0$ is distributed as the variable $\min(S, I)$ and its mean value is:

$$\int_{0}^{\infty} y P\{y < \min(S, I) < y + dy\} = \{\lambda + \mu\}^{-1}$$

So one obtains the recurrence:

$$(\lambda + \mu)T(n) = \mu T(n+1) + \lambda T(n-1), \quad n > 0 \tag{1.3.4}$$

of which the solution is:

$$T(n) = \left(\frac{\lambda}{\mu}\right)^{n} T(0). \tag{1.3.5}$$

But $m(0) = 1$ from the choice which we have made for the interval $[a, b]$, and

T(0) will be quite simply the mean time spent in state 0 in that interval. We shall exploit for one last time the memoryless property which implies that:

$$T(0) = E[1] = \frac{1}{\lambda}$$

Finally we have:

$$T(n) = \frac{1}{\lambda} \left(\frac{\lambda}{\mu} \right)^n \qquad (1.3.6)$$

and T, the mean duration of the interval in question, will be:

$$\begin{cases} T = E[b-a] = \frac{1}{\lambda} \sum_{n=0}^{\infty} \left(\frac{\lambda}{\mu} \right)^n = \frac{\lambda^{-1}}{1 - \lambda/\mu} & \text{if} \quad \lambda/\mu < 1, \\ T = \infty & \text{if} \quad \lambda/\mu \geqslant 1. \end{cases}$$

The *probability of state* n is now defined by:

$$p(n) = \frac{T(n)}{T}$$

which is valid only if $T < \infty$, that is if $\lambda/\mu < 1$. So we have:

$$p(n) = \left(\frac{\lambda}{\mu} \right)^n \left(1 - \frac{\lambda}{\mu} \right). \qquad (1.3.7)$$

In fact, the interval $[a, b]$ which we have chosen in this paragraph allows us to characterize completely the process $N(t)$ for *all values* of t and not only for $a < t < b$. To convince ourselves, let us examine all the events which occur for $t > b$. The process $N(t)$ for $t > b$ will be determined by the duration of the first service, and by the time t' of the first arrival which occurs after b.

$(t' - b)$ is a time between two successive arrivals and so does not depend on $N(\tau)$ for $\tau < b$. Similarly, at time b a new service starts and its duration no longer depends on previous events. So it can be deduced that $N(t)$ for $t > b$ is independent of $N(\tau)$ for $\tau < b$: one says that $t = b$ is a *regeneration point* of the process $N(t)$. If there exists an infinite series of successive instants $b_1, b_2, b_3, \ldots$ at which the queue passes from the empty state to the state $n = 1$, it is evident that $N(t)$ taken over an interval $b_i < t < b_{i+1}$ will be independent of $N(t)$ over other intervals. But since successive interarrivals and services are identically distributed, one can say that for all $i \neq j$, $\tau \geqslant 0$,

$$P\{N(\tau + b_i), b_i + \tau < b_{i+1}\} = P\{N(\tau + b_j), b_j + \tau < b_{j+1}\},$$

that is the process $N(t)$ repeats itself in successive intervals $[b_i, b_{i+1}]$. A process which has this property is called a *regenerative process*.

The practical conclusion is that the results which we have obtained for the interval $[a, b]$ are in fact the same for all intervals $[b_i, b_{i+1}]$. This infinite series of instants b_i, $i > 1$, will exist if $E[b - a] < \infty$, that is if $\lambda/\mu < 1$. It can also be proved, but more advanced mathematics would be required, that, almost surely (see Cinlar [1]):

$$p(n) = \frac{T(n)}{E[b-a]} = \lim_{\tau \to \infty} \frac{1}{\tau} \int_0^{\tau} dt \, 1(N_t = n) \qquad (1.3.8)$$

where

$$1(N_t = n) = \begin{cases} 1 & \text{if } N_t = n, \\ 0 & \text{otherwise.} \end{cases}$$

This formula indicates that $p(n)$ can be interpreted also as being, asymptotically, the proportion of time spent in state n (almost certainly).

1.3.3. Differential equation approach and the Chapman–Kolmogorov equation

The results obtained in Section 1.3.2 can equally be deduced from a study of 'dynamic' behaviour. This approach resembles that which a physicist would use to study the evolution of the state $N(t)$ of the queue as a function of time.

Consider an instant t for which $N(t) = j > 0$ and let us examine what can happen in a very short period of time Δt. One arrival will occur with a probability:

$$P\{I - y < \Delta t | I > y\} = P\{y < I < y + \Delta t\} P\{I > y\},$$
$$= \frac{e^{-\lambda y} - e^{-\lambda(y + \Delta t)}}{e^{-\lambda y}},$$
$$\approx \lambda \Delta t + o(\Delta t),$$

if the last arrival occurred at the instant $t - y$. A service will finish with the probability:

$$P\{S - x < \Delta t | S > x\} = \mu \Delta t + o(\Delta),$$

if the last instant of end of service is $t - x$. Certainly, thanks to the memoryless property of the exponential distribution, x and y have no influence on these probabilities. Therefore we have the following changes of state:

$$N(t) = j \xrightarrow{(1 - \mu \Delta t)\lambda \Delta t} N(t + \Delta t) = j + 1,$$

$$N(t) = j \xrightarrow{(1 - \lambda \Delta t)\mu \Delta t} N(t + \Delta t) = j - 1,$$

$$N(t) = j \xrightarrow{(1 - \lambda \Delta t)(1 - \mu \Delta t)} N(t + \Delta t) = j$$

where it is appropriate to add $o(\Delta t)$ to each of the transition probabilities.

We can define the probability:

$$p(j, t) = P\{N(t) = j\}$$

which must satisfy the equation:

$$p(j, t + \Delta t) = \lambda \Delta t(1 - \mu \Delta t)p(j - 1, t)$$
$$+ \mu \Delta t(1 - \lambda \Delta t)p(j + 1, t) \tag{1.3.9}$$
$$+ (1 - \lambda \Delta t)(1 - \mu \Delta t)p(j, t) \quad \text{if } j < 0$$

and the equation:

$$p(0, t + \Delta t) = (1 - \lambda \Delta t)p(0, t)$$
$$+ \mu \Delta t(1 - \lambda \Delta t)p(1, t) \quad \text{if } j = 0 \tag{1.3.10}$$

Putting:

$$\frac{d}{dt}p(j, t) = \lim_{\Delta t \to 0} [p(j, t + \Delta t) - p(j, t)]/\Delta t$$

and using the fact that

$$\lim_{\Delta t \to 0} \frac{o(\Delta t)}{\Delta t} = 0,$$

we obtain:

$$\frac{d}{dt}p(j, t) = \lambda p(j - 1, t) + \mu p(j + 1, t) - (\lambda + \mu)p(j, t) \quad \text{for } j > 0 \tag{1.3.11}$$

and

$$\frac{d}{dt}p(0, t) = \mu p(1, t) - \lambda p(0, t). \tag{1.3.12}$$

The solution for the equilibrium state (or the stationary state), if it exists, must satisfy:

$$\lim_{t \to \infty} \frac{d}{dt}p(j, t) = 0$$

and if we let

$$p(j) = \lim_{t \to \infty} p(j, t),$$

this leads to the equilibrium equations which we encountered in the preceding paragraph:

$$p(j)(\lambda + \mu) = \lambda p(j - 1) + \mu p(j + 1), \quad j > 0,$$
$$\lambda p(0) = \mu p(1)$$

for which the solution will be as we have already established:

$$p(j) = \left(\frac{\lambda}{\mu}\right)^j \left(1 - \frac{\lambda}{\mu}\right), \quad j \geq 0$$

if

$$\lambda/\mu < 1.$$

1.3.4. The Poisson process

Equations (1.3.11) and (1.3.12) allow us to study the process which simply governs the arrival of customers in the queue: it is sufficient to suppose that the services do not occur ($\mu = 0$) so that the number N(t) of customers in the queue becomes the number A(t) of customers who have joined the queue up to time t. We denote the probability that the number of arrivals between times 0 and t is n by $\pi(n, t)$:

$$\pi(n, t) = P\{A(t) = n\}$$

which must satisfy equations (1.3.11) and (1.3.12) with $\mu = 0$ (the 'rate of service' is zero, so the 'time of service' is infinite). A(t) is a non-decreasing random process with 'jumps' (the arrivals, since there are no services).

$$\frac{d}{dt} \pi(n, t) = \lambda \pi(n - 1, t) - \lambda \pi(n, t), \quad n > 0,$$

$$\frac{d}{dt} \pi(0, t) = - \lambda \pi(0, t).$$

The second equation gives immediately:

$$\pi(0, t) = e^{-\lambda t},$$

if we assume A(0) = 0, then $\pi(0, 0) = 1$.

Let us write the generating function:

$$\tilde{\pi}(x, t) = \sum_{n=0}^{\infty} \pi(n, t) x^n, \quad \text{for} \quad |x| \leqslant 1.$$

Then the preceding equations give:

$$\frac{d}{dt} \tilde{\pi}(x, t) = \lambda x \tilde{\pi}(x, t) - \lambda \tilde{\pi}(x, t)$$

$$= - \lambda(1 - x) \tilde{\pi}(x, t)$$

of which the solution is:

$$\tilde{\pi}(x, t) = \alpha(x) e^{-\lambda(1 - x)t}.$$

But we know that:

$$\tilde{\pi}(x, 0) = \pi(0, 0) = 1$$

since A(0) = 0. Hence:

$$\tilde{\pi}(x, t) = e^{-\lambda(1 - x)t}$$

which can equally be written:

$$\tilde{\pi}(x, t) = \sum_{n=0}^{\infty} e^{-\lambda t} \frac{(\lambda t x)^n}{n!}.$$

We conclude that for $n \geqslant 0$:

$$\pi(n, t) = P\{A(t) = n\} = e^{-\lambda t}\frac{(\lambda t)^n}{n!}.$$

The process $A(t)$, $t \geqslant 0$, defined in this way is known as a Poisson process and the law $\pi(n, t)$ is the Poisson law with parameter (λt).

We notice that the Poisson process is quite simply defined by a sequence of events (the 'arrivals') such that the times between successive arrivals constitute the independent exponential random variables.

1.4 - GENERAL QUEUES WITH A SINGLE SERVER

In this paragraph we extend the class of queues which we have studied in the preceding paragraph to allow the study of systems in which the service times or the interarrivals are not exponential. We shall start by analysing a system for which we shall make no restrictive assumptions: this will allow us to obtain a perfectly general and extremely useful result which is called *Little's formula*. We shall also show a relation concerning the length of the queue measured at the instants of arrival and departure.

Some particular analytic results will then be obtained by making certain restrictive assumptions. Several applications of the results of Sections 1.3 and 1.4 are presented in Section 1.5.

1.4.1. Little's formula

As in the preceding section, let $N(t)$ be the number of customers in the system (those which are in the queue, plus the customer being served). $A(t)$ is the number of customers who have arrived up to the instant t:

$$A(t) = \sup_{a_i < t} i$$

assuming that $A(0) = 0$. Let d_i, $i \geqslant 1$, be the departure time of the ith customer and $D(t)$ be the number of departures which have occurred up to time t:

$$D(t) = \sup\{i : d_i < t\}.$$

Therefore we have:

$$N(t) = A(t) - D(t).$$

The *response time* of the ith customer is:

$$T_i = d_i - a_i.$$

This is the total time which he has spent in the system. His *waiting time* W_i is:

$$W_i = T_i - S_i,$$

that is the time spent in the system less the service time.

12

As in Section 1.3.2, let us consider two instants a and b $(a < b)$ such that $N(a) = N(b) = 0$, with $N(a^+) = N(b^+) = 1$.

Let us put:

$$\bar{N} = \int_a^b N(t)\,dt/(b-a)$$

the mean number of customers in the system in the interval $[a, b]$. Let us put:

$$a = a_{i_1} < a_{i_2} < \cdots < a_{i_k} < b$$

the instants at which arrivals occur in the interval $[a, b]$, an arrival occurring at the instant b being excluded. That is we suppose that k arrivals have occurred in the interval. If d_{ij} is the instant of departure of a customer arriving at the instant a_{ij}, the mean response time of these customers will be:

$$\bar{T} = \sum_{j=1}^k (d_{i_j} - a_{i_j})/k$$

where $d_{i_k} \leqslant b$ since $N(b) = 0$. If we put:

$$\lambda(a, b) = k/(b-a),$$

we can write:

$$\lambda(a, b)\bar{T} = \sum_{j=1}^k (d_{i_j} - a_{i_j})/(b-a).$$

But:

$$\int_a^b N(t)\,dt = \int_a^b (A(t) - D(t))\,dt = \sum_1^k (d_{i_j} - a_{i_j}).$$

By way of example, it is clear that in Figure 1.4

$$\int_a^b N(t)\,dt = (d_{i_4} - a_{i_1}) + (d_{i_3} - a_{i_2}) + (d_{i_2} - a_{i_4}) + (d_{i_1} - a_{i_3}).$$

This gives us *Little's formula* for the deterministic case:

$$\lambda(a, b)\bar{T} = \bar{N}. \tag{1.4.1}$$

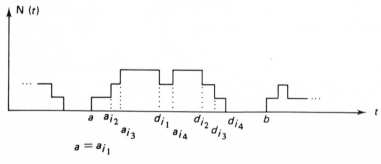

Figure 1.4 Illustration of Little's formula

1.4.2. Arrival and departure instants

An interesting symmetry is established between the instants of arrival at, and departure from, the queue which involves $N(t)$.

We shall assume that:

−there exists an infinite series of instants $0 < b_1 < b_2 < \cdots$ such that:

$$N(b_i) = 0 \quad \text{and} \quad N(b_i^+) = 1, \quad i \geq 1, \tag{1.4.2}$$

as in the preceding sections, and that:
−arrivals and departures of customers occur singly (there are no group arrivals or departures).

Let $N(a_i)$ be the number of customers in the system just *before* the arrival of the ith customer and $N(d_j^+)$ the number of them in the system just *after* the jth departure. For each a_{il} such that $b_i \leq a_{il} < b_{i+1}$, there exists a d_{i_m}, such that $b_i < d_{i_m} < b_{i+1}$, and:

$$N(a_{il}) = N(d_{i_m}^+). \tag{1.4.3}$$

On the other hand, the sequence of arrival and departure times taken in this interval can be decomposed into distinct pairs (a_{il}, d_{i_m}) having this property.

Let us define for the interval $[b_i, b_{i+1}]$, containing k_i arrivals and departures:

$$\bar{N}_a(n) = \left[\sum_{l=1}^{k_i} 1(N(a_{il}) = n) \right] \Big/ k_i,$$

$$\bar{N}_d(n) = \left[\sum_{l=1}^{k_i} 1(N(d_{il}^+) = n) \right] \Big/ k_i$$

where the function

$$1(N(a_{il}) = n) = \begin{cases} 1 & \text{if } N(a_{il}) = n, \\ 0 & \text{otherwise}, \end{cases}$$

the proportion of arrival times (or departure times respectively) finding (or leaving respectively) the system with n customers in the queue. Equation (1.4.3) has the immediate consequence that for all $n \geq 0$, and all intervals $[b_i, b_{i+1}]$,

$$\bar{N}_a(n) = \bar{N}_d(n). \tag{1.4.4}$$

So all estimates of the state of the system just before arrivals become equivalent to estimates just *after* departures.

1.4.3. Extension of some deterministic results to the case of the regenerative process $N(t)$: Little's formula

In this section, we shall study $N(t)$ supposing that:

−the interarrivals $I_1, I_2, \ldots$ are independent and identically distributed (i.i.d), and that:
−successive services $S_1, S_2, \ldots$ are also i.i.d. and independent of the interarrivals.

We shall assume that an infinite sequence of instants

$$0 < b_1 < b_2, \ldots$$

exists such that for each $i \geqslant 1$:

$$N(b_i) = 0, \quad N(b_i^+) = 1.$$

It is clear that the random process $\{N(t), t > b_i\}$ is independent of $\{N(t), t \leqslant b_i\}$ for each $i \geqslant 1$ since the interarrivals and services consist of series of independent variables.

The fact that each of these sequences consists of identically distributed variables results in the process $\{N(t + b_i), t \geqslant 0\}$ and the process $\{N(t + b_j), t \geqslant 0\}$ being identical for all i and j. One says that the b_i are *points of regeneration* of $N(t)$, and that this is a *regenerative process*.

This type of process has been studied in detail (cf. Cinlar [1]) and the following result is often used.

Theorem

Let f be a (measurable) function; then for any $i \geqslant 1$,

$$\lim_{t \to \infty} \frac{\int_0^t f(N(t)) \, dt}{t} = E\left[\int_{b_i}^{b_{i+1}} f(N(t)) \, dt \right] \Big/ E[b_{i+1} - b_i].$$

This theorem indicates that it is sufficient to calculate certain interesting quantities concerning $N(t)$ over an *any interval* $[b_i, b_{i+1}]$ in order to obtain this same quantity for the whole process over $[0, \infty]$. We shall apply this result principally to obtain Little's formula (1.4.1) for a regenerative process $N(t)$.

Applying it directly to $N(t)$, we obtain:

$$E[N] \equiv \lim_{t \to \infty} \frac{\int_0^t N(t) \, dt}{t},$$

$$= E\left[\int_{b_i}^{b_{i+1}} N(t) \, dt \right] \Big/ E[b_{i+1} - b_i].$$

Let us put $\lambda \equiv E[I_i]^{-1}$ and K_i the number of customers entering the system from the initial instant up until time b_i; it is clear that one can write:

$$\lambda = E[K_{i+1} - K_i]/E[b_{i+1} - b_i],$$

if $K_{i+1} - K_i$ is the number of arrivals (and hence also departures) occurring in the interval $[b_i, b_{i+1}]$. This gives us:

$$E[N]/\lambda = E\left[\int_{b_i}^{b_{i+1}} N(t) \, dt \right] \Big/ E[K_{i+1} - K_i]. \tag{1.4.5}$$

Let us now examine the sequence of the *waiting times* $W_1, W_2, \ldots$, introduced in Section 1.4.1. We notice that the waiting times of customers arriving at times $0, b_1, b_2, \ldots$, are *zero*, and that the other waiting times are not:

$$W_1 = 0, \quad W_2 > 0, \ldots, W_{K_1+1} = 0, \quad W_{K_1+2} > 0, \ldots, \quad W_{K_2} > 0,$$
$$W_{K_2+1} = 0, \quad W_{K_2+2} > 0, \ldots.$$

One can also easily establish the recurrence for $i \geq 1$:

$$W_{i+1} = \begin{cases} 0 & \text{if } W_i + S_i - I_{i+1} < 0, \\ W_i + S_i - I_{i+1} & \text{otherwise.} \end{cases} \tag{1.4.6}$$

The sequence W_i thus defined is a random process in which the integers $1, 2, \ldots$, play the role of 'times'. It is a regenerative process of which the 'instants' of regeneration are $1, K_1 + 1, K_2 + 1, \ldots$ This is easy to verify since W_{K_i+j} will have exactly the same distribution as W_{1+j} for all $i \geq 1, j \geq 0$. So by using a slightly different form of the above theorem, one obtains:

$$E[W] = \lim_{k \to \infty} \frac{\sum_1^k W_i}{k} = \frac{E\left[\sum_{K_i+1}^{K_{i+1}} W_j\right]}{E[K_{i+1} - K_i]}.$$

We are also interested in the response time:

$$T_i = W_i + S_i$$

and in

$$E[T] = \lim_{k \to \infty} \frac{\sum_1^k (W_i + S_i)}{k}.$$

The process of the T_i will have the same regenerative structure as W_i, since the S_i are i.i.d. Hence:

$$E[T] = \frac{E\left[\sum_{K_i+1}^{K_{i+1}} T_j\right]}{E[K_{i+1} - K_i]}.$$

But in Section 1.4.1, it is shown that:

$$\sum_{K_i+1}^{K_{i+1}} T_j = \int_{b_i}^{b_{i+1}} N(t)\,dt.$$

Hence

$$E[T] = E\left[\int_{b_i}^{b_{i+1}} N(t)\,dt\right] \bigg/ E[K_{i+1} - K_i], \tag{1.4.7}$$

$$= E[N]/\lambda$$

which is Little's formula for the random case.

1.4.4. Probability of finding the queue empty

Another equally general result which can be deduced concerns the asymptotic probability that the queue is empty. We continue the assumptions of Section 1.4.3: the I_i and S_i consist of mutually independent series of i.i.d. variables. This probability is defined by

$$p(0) \equiv \lim_{t \to \infty} \frac{\displaystyle\int_0^t 1(N(t) = 0)\,dt}{t}. \tag{1.4.8}$$

We again suppose the existence of an already defined infinite series of instants $0 < b_1 < b_2 < \cdots$ which constitute the regeneration points of the process $N(t)$. Since $1(.)$ used in equality (1.4.8) is a measurable function, one can apply the theorem concerning regenerative processes used in Section 1.4.3, to show that for all $i \geqslant 1$:

$$p(0) = \frac{E\left[\displaystyle\int_{b_i}^{b_{i+1}} 1(N(t) = 0)\,dt\right]}{E[b_{i+1} - b_i]}$$

or

$$p(0) = 1 - \frac{E\left[\displaystyle\int_{b_i}^{b_{i+1}} 1(N(t) > 0)\,dt\right]}{E[b_{i+1} - b_i]}.$$

Let k be the number of arrivals (or departures) in the interval $[b_i, b_{i+1}]$, and let $I_{i_1}, \ldots, I_{i_k}$ be the interarrivals corresponding to this interval:

$$b_{i+1} = \sum_{j=1}^{k} I_{i_j} + b_i.$$

Since:

$$\int_{b_i}^{b_{i+1}} 1(N(t) > 0)\,dt = \sum_{j=1}^{k} S_{i_j}$$

where $S_{i_1}, \ldots, S_{i_k}$ are the service times occuring in the interval (S_{i_1} is the service time which starts at the instant b_i), we obtain (*):

$$E[b_{i+1} - b_i] = E[k]E[I_j],$$

$$E\left[\int_{b_i}^{b_{i+1}} 1(N(t) > 0)\,dt\right] = E[k]E[S_j]$$

and hence

$$\rho(0) = 1 - \frac{E[S_j]}{E[I_j]}.$$

If one writes $\lambda \equiv (E[I_j])^{-1}$ and $\mu \equiv (E[S_j])^{-1}$, one obtains the classic formula:

$$p(0) = 1 - \frac{\lambda}{\mu}.$$

(*) Here we are using Wald's lemma [9].

1.4.5. Kendall's notation

Kendall's notation is the following:

$$A/B/C/K/m/Z.$$

It specifies the six factors which must be known in order to define a queue: arrival process/service process/number of servers/maximum capacity/population of users/service discipline.

In the description of the arrival process or the service process, the following symbols are used:

GI : general independent distribution
G : general distribution,
H_k : hyperexponential distribution of order k,
E_k : Erlang distribution
M : exponential distribution
D : constant distribution

The principal service disciplines used are the following:

FIFO : first come, first served (first in, first out),
LCFS : last come, first served,
FIRO : first in random out.

When the last three elements of Kendall's notation are not specified, it is understood that $Z = FIFO$, $m = +\infty$ and $K = +\infty$.

Example The queue $M/E_k/2/2/6/FIRO$ has Poisson arrivals, an Erlang-k service distribution, two servers, a maximum capacity of two customers; only six users can occupy this queue and the service discipline is random.

1.5 - M/GI/1 QUEUE

We shall consider, in this section, a queue having a Poisson arrival process and any service time distribution, with a FIFO service discipline.

To study this queue, the simplest approach is to construct the 'embedded' Markov chain. That is what we shall do first.

Let $A(t)$ be the number of arrivals in the queue in the interval $[0, t]$ and $N(t)$ the number of customers in the system at time t. To simplify the notation we shall suppose that $N(0) = A(0) = 0$. Again let $S_1, S_2, \ldots, S_n, \ldots$ be the service times of customers $1, 2, \ldots, n, \ldots$ We shall suppose that the process $\{A(t), t \geqslant 0\}$ is a Poisson process with parameter λ. The service times S_n are independent of each other and of the arrival process; they are identically distributed following the probability distribution $G(s)$ with a mean μ^{-1}. We wish to find the stability condition of this queue as well as the distribution of the number of customers in the equilibrium state.

Let X_n be the number of customers in the system just after the nth departure from the queue. We have:

Theorem

The process $\{X_n\}_{n \geqslant 1}$ is a Markov chain with a transition matrix P. Let:

$$p_k = \int_0^\infty \frac{e^{-\lambda s}(\lambda s)^k}{k!} dG(s), \quad k = 0, 1, \ldots$$

then

$$P = \begin{bmatrix} p_0 & p_1 & p_2 & p_3 & \cdot & \cdot & \cdot \\ p_0 & p_1 & p_2 & p_3 & \cdot & \cdot & \cdot \\ 0 & p_0 & p_1 & p_2 & p_3 & \cdot & \cdot \\ 0 & 0 & p_0 & p_1 & \cdot & \cdot & \cdot \\ 0 & \cdot & 0 & p_0 & \cdot & \cdot & \cdot \\ & \cdot & \cdot & 0 & \cdot & \cdot & \cdot \\ \cdot & \cdot & \cdot & \cdot & 0 & \cdot & \cdot \\ \cdot & & & & & \cdot & \cdot \\ 0 & & & & \cdot & \cdot & \cdot \end{bmatrix}$$

Proof One will have shown that $\{X_n\}_{n \geqslant 1}$ is a Markov chain if one proves that:

$$P\{X_{n+1} = i | X_0, X_1, \ldots, X_n\} = P\{X_{n+1} = i | X_n\} \tag{1.5.1}$$

Let d_n and a_n be the instants of departure and arrival respectively of the nth customer in the system. If at the instant d_n the number of customers in the queue is not zero, we have:

$$X_{n+1} = X_n + A(d_{n+1}) - A(d_n) - 1, \quad \text{or} \quad d_{n+1} = d_n + S_{n+1}. \tag{1.5.2}$$

In contrast, if $X_n = 0$, we simply have:

$$X_{n+1} = A(a_{n+1} + S_{n+1}) - A(a_{n+1}). \tag{1.5.3}$$

As the arrival process is Poisson, the number of arrivals in an interval of duration S_{n+1} depends only on S_{n+1}. In the two preceding cases (1.5.2) and (1.5.3), relation (1.5.1) is verified: the chain X_n is a Markov chain.

Let us now calculate the elements of the transition matrix. The first row of the matrix is determined by $P\{X_{n+1} = i | X_n = 0\}$. From (1.5.3) we have:

$$P\{X_{n+1} = i | X_n = 0\} = P\{A(a_{n+1} + S_{n+1}) - A(a_{n+1}) = i\}$$

$$= P\{A(S_{n+1}) = i\}, = \int_0^\infty \frac{e^{-\lambda s}(\lambda s)^i}{i!} dG(s) = p_i.$$

The elements of the jth row of the transition matrix are determined by $P\{X_{n+1} = i | X_n = j\}$. From relation (1.5.2) we obtain:

$$P\{X_{n+1} = i | X_n = j\} = P\{A(d_n + S_{n+1}) - A(d_n) - 1 = i - j\},$$
$$= P\{A(S_{n+1}) = i - j + 1\}$$
$$= p_{i-j+1} \quad \text{if } i \geqslant j - 1,$$
$$= 0 \qquad \text{otherwise.}$$

We have determined the entire transition matrix. We are now able to approach the problem of the existence of a stationary solution for the M/GI/1 queue. Firstly, notice from its transition matrix, that the Markov chain $\{X_n\}$ is irreducible and aperiodic.

Theorem

The chain $\{X_n\}$ is positive recurrent if and only if

$$\rho = \frac{\lambda}{\mu} < 1.$$

To prove this result we shall use the following classic result: an irreducible and aperiodic Markov chain $\{X_n\}$ is positive recurrent if and only if the linear systems $\Pi P = \Pi$ and $\Pi 1 = 1$ have one unique, strictly positive, solution. Let us put:

$$\Pi = (\Pi_0, \Pi_1, \ldots).$$

The two linear systems $\Pi P = \Pi$ and $\Pi 1 = 1$ are written in the following manner:

$$\begin{cases} \Pi_0 = \Pi_0 p_0 + \Pi_1 p_0, \\ \Pi_1 = \Pi_0 p_1 + \Pi_1 p_1 + \Pi_2 p_0, \\ \Pi_2 = \Pi_0 p_2 + \Pi_1 p_2 + \Pi_2 p_1 + \Pi_3 p_0, \\ \ldots\ldots\ldots\ldots \end{cases} \tag{1.5.4}$$

and

$$\Pi_0 + \Pi_1 + \Pi_2 + \cdots = 1.$$

Let us put:

$$r_k = 1 - p_0 - p_1 - \cdots - p_k.$$

From the linear system (1.5.4), by adding the equations term by term, we obtain:

$$\begin{cases} \Pi_1 p_0 = \Pi_0 r_0, \\ \Pi_2 p_0 = \Pi_0 r_1 + \Pi_1 r_1, \\ \Pi_3 p_0 = \Pi_0 r_2 + \Pi_1 r_2 + \Pi_2 r_1, \\ \ldots\ldots\ldots\ldots \end{cases} \tag{1.5.5}$$

For all $\Pi_0 \geqslant 0$, the system (1.5.5) has a unique solution with the exception of a multiplicative constant. Let us add the relations of system (1.5.5) term by term:

$$p_0 \left(\sum_{i=1}^{\infty} \Pi_i \right) = \Pi_0 (r_0 + r_1 + \cdots) + (r_1 + r_2 + \cdots) \sum_{i=1}^{\infty} \Pi_i.$$

Notice that $r_0 + r_1 + \cdots + r_n$ is the mean number of arrivals during a service, hence $r_0 + r_1 + \cdots + r_n = \rho$. This yields

$$(1 - r_0) \sum_{i=1}^{\infty} \Pi_i = \Pi_0 \rho + (\rho - r_0) \sum_{i=1}^{\infty} \Pi_i. \tag{1.5.6}$$

If $\rho < 1$, relation (1.5.6) can be written in the form:

$$\sum_{i=1}^{\infty} \Pi_i = \frac{\rho}{1 - \rho} \Pi_0.$$

The linear system $\Pi 1 = 1$ has a unique strictly positive solution if $\Pi_0 = 1 - \rho$.

Hence if $\rho < 1$, the chain $\{X_n\}$ is positive recurrent. The stationary solution of the M/GI/1 queue is obtained from $\Pi_0 = 1 - \rho$ and from system (1.5.5). Conversely, if $\rho \geqslant 1$, the equality (1.5.6) implies:

$$\sum_{i=1}^{\infty} \Pi_1 = 0 \quad \text{if } \Pi_0 = 0 \quad \text{or} \quad \sum_{i=1}^{\infty} \Pi_i = +\infty \quad \text{if } \Pi_0 > 0.$$

If $\rho \geqslant 1$, the equation $\Pi 1 = 1$ has no solution and the chain $\{X_n\}$ is not positive recurrent. The M/GI/1 queue cannot have a stationary solution.

We have studied the M/GI/1 queue by way of the Markov chain embedded at the departure times. We have obtained a stationary solution at these particular instants. We must now study arbitrary instants. We shall show in the following paragraphs that the stationary solution obtained at any instant of the queueing process is identical to that of the embedded Markov chain, that is $p(k) = \Pi_k$, $k = 0, 1, 2, \ldots$.

Recall the previous definition of a Markov renewal process. Let $X_n, n = 1, 2, \ldots$, be a series of random variables taking their values in a countable set E and let $T_n, n = 1, 2, \ldots$ be a sequence of increasing random variables having values in R_+.

The stochastic process $(X, T) = \{X_n, T_n; n \in N\}$ is a Markov renewal process in the space E if

$$P\{X_{n+1} = j, T_{n+1} - T_n \leqslant t | X_0, \ldots, X_n; T_0, \ldots, T_n\}$$
$$= P\{X_{n+1} = j, T_{n+1} - T_n \leqslant t | X_n\},$$

for all $n \in N$, $j \in E$ and $t \in R_+$.

The stochastic process (X, T) is homogeneous in time if, for all $i, j \in E$ and $t \in R_+$, the quantity

$$P\{X_{n+1} = j, T_{n+1} - T_n \leqslant t | X_n = i\} = P(i, j, t)$$

is independent of n. The probabilities $P(i, j, t)$ form the semi-Markovian kernel associated with (X, T).

Theorem

The process $\{X_n, d_n\}$ is a Markov renewal process with transition matrix $P(t)$ (semi-Markovian kernel) given by:

$$P(t) = \begin{bmatrix} q_0(t) & q_1(t) & q_2(t) & \cdots \\ p_0(t) & p_1(t) & p_2(t) & \cdots \\ (0) & p_0(t) & p_1(t) & \cdots \end{bmatrix}$$

where

$$p_n(t) = \int_0^t \frac{e^{-\lambda s}(\lambda s)^n}{n!} \, dG(s), \quad n = 0, 1, 2, \ldots,$$

and

$$q_n(t) = \int_0^t \lambda e^{-\lambda s} p_n(t-s) \, ds, \quad n = 0, 1, 2, \ldots.$$

Proof The stochastic process $\{X_n, d_n\}$ is a Markov renewal process if:

$$P\{X_{n+1} = j, d_{n+1} - d_n \leqslant t | X_0, \ldots, X_n; d_0, \ldots, d_n\}$$
$$= P\{X_{n+1} = j, d_{n+1} - d_n \leqslant t | X_n\} \tag{1.5.7}$$

for all n and positive integer j and $t \in R_+$.

As the arrival process is Poisson and the service times are independent of each other, relation (1.5.7) is verified.

The elements of the $(i+1)$th row of the transition matrix are determined by (row 0 is determined later):

$$P(i, j, t) = P\{X_{n+1} = j, \quad d_{n+1} - d_n \leqslant t | X_n = i\},$$
$$= P\{A(S_{n+1}) = j - i + 1, \quad d_{n+1} - d_n \leqslant t\}$$
$$= p_{j-i+1}(t).$$

The first row of the transition matrix is determined by the terms:

$$P(0, j, t) = P\{X_{n+1} = j, \quad d_{n+1} - d_n \leqslant t | X_n = 0\},$$
$$= \int_0^t P\{a_{n+1} - d_n = s\} P\{A(S_{n+1}) = j, d_{n+1} - a_{n+1} \leqslant t - s\} \, ds,$$
$$= q_j(t).$$

Notice that the transition matrix of the embedded chain $\{X_n\}$ is obtained with $P(i, j, \infty)$.

We are going to use the fundamental theorem of Markov renewal processes, which allows the stationary solution at any instant to be related to the stationary solution of the embedded Markov chain.

The theorem is written in the following manner (see Cinlar [9]):

$$\lim_{t \to \infty} P\{N(t) = k\} = p(k) = \sum_j \frac{\Pi_j}{m} \int_0^\infty A(j, k, t) \, dt, \tag{1.5.8}$$

where Π_j is the stationary probability of the embedded chain (which we have already calculated), m is the mean time between two instants of the embedded Markov chain and $A(j, k, t)$ is the probability that the system changes from state j to state k during the time interval $[0,t]$. $A(j,k,t)$ is the kernel of the Markov renewal process associated with $P(j,k,t)$. We have:

$$A(j,k,t) = \begin{cases} e^{-\lambda t} & \text{if } j = k = 0, \\[2mm] \int_0^t \lambda e^{-\lambda(t-s)} \dfrac{e^{-\lambda s}(\lambda s)^{k-1}}{(k-1)!}(1 - G(s))\,ds & \text{if } k > j = 0, \\[2mm] [1 - G(t)]\dfrac{e^{-\lambda t}(\lambda t)^{k-j}}{(k-j)!} & \text{if } k \geqslant j > 0, \\[2mm] 0 & \text{otherwise.} \end{cases}$$

The case $j = k = 0$ represents the case where no arrivals occur during time t. If $k > j = 0$, during time $t - s$, there are no arrivals, the first arrival occurs at time $t - s$ and the $k - 1$ other arrivals occur during the remaining time s, which is equal to the service time of the customer who arrived at time $t - s$. This occurs with the probability $1 - \int_0^s g(u)\,du = 1 - G(s)$. If $k \geqslant j > 0$ at the time origin, the server is active and $k - j$ arrivals occur during the service epoch.

Let us return to the fundamental theorem: if the embedded Markov chain is positive recurrent (that is if $\rho < 1$), the Markov renewal process is ergodic and the mean time between two instants of the embedded Markov chain is simply λ^{-1}. As we wish to show that $\lim_{t \to \infty} P\{N(t) = k\} = \Pi_k$, it is sufficient to verify the equality obtained from (1.5.8), that is:

$$\Pi_k = \sum_j \lambda \Pi_j \int_0^\infty A(j,k,t)\,dt. \tag{1.5.9}$$

Let v_k be the term on the right of equation (1.5.9). To prove this identity, it is sufficient to verify the equality of the series $V(z) = \sum_{k=0}^\infty v_k z^k$ and $U(z) = \sum_{k=0}^\infty \Pi_k z^k$ for all $z \in [0,1]$. Set $\alpha = \lambda(1 - z)$, we obtain from (1.5.9):

$$\begin{aligned} (1 - z)V(z) &= \alpha \sum_{j=0}^\infty \Pi_j \int_0^\infty dt \sum_{k=j}^\infty A(j,k,t)z^k, \\ &= \Pi_0 \left\{ \int_0^\infty \alpha e^{-\lambda t}\,dt \right. \\ &\quad \left. + z \int_0^\infty dt \int_0^t \alpha e^{-\alpha s}[1 - G(s)]\lambda e^{-\lambda(t-s)}\,ds \right\} \\ &\quad + \sum_{j=1}^\infty \Pi_j z j \int_0^\infty \alpha e^{-\alpha t}[1 - G(t)]\,dt, \\ &= \Pi_0[1 - z + z(1 - F(z))] + \sum_{j=1}^\infty \Pi_j z j(1 - F(z)), \\ &= U(z)[1 - F(z)] + \Pi(0)(1 - z)F(z) \end{aligned} \tag{1.5.10}$$

where

$$F(z) = \int_0^\infty e^{-\alpha t} g(t)\,dt. \quad F(z) = \int_0^\infty \alpha e^{-\alpha t} G(t)\,dt,$$

$$= \int_0^\infty e^{-\alpha t} g(t)\,dt.$$

Using the system $\Pi = \Pi P$, we obtain:

$$zU(z) = \sum_i \Pi_i \sum_j P(i,j) z^{j+1},$$
$$= [\Pi_0 z + \Pi_1 z + \Pi_2 z^2 + \cdots] F(z), \qquad (1.5.11)$$
$$= U(z)F(z) - \Pi_0(1-z)F(z).$$

From (1.5.10) and (1.5.11) by eliminating $\Pi_0(1-z)F(z)$, we obtain

$$(1-z)V(z) = U(z)[1 - F(z)] + G(z)F(z) - zG(z) = (1-z)U(z).$$

This shows the equality for $z < 1$.

We have shown that the stationary solution, at any instant, is the same as that obtained for the instants of the embedded chain $\{X_n\}$. In particular, the mean length of the queue is given by $E[X_n]$ which we shall now calculate.

The Pollaczek–Khintchine formula

From relations (1.5.2) and (1.5.3), we can write a simple relation between X_{n+1} and X_n:

$$X_{n+1} = X_n + B_n - C_n \qquad (1.5.12)$$

where $C_n = 1 - P\{X_n = 0\}$, and B_n is the number of arrivals during time T_n.

In the case where the chain $\{X_n\}$ is positive recurrent, that is $\rho < 1$, we have:

$$\lim_{n \to \infty} E[C_n] = 1 - \lim_{n \to \infty} P\{X_n = 0\} = 1 - \Pi_0 = \rho.$$

We have also:

$$E[B_n] = \rho,$$
$$E[B_n^2] = \rho + \lambda^2 \int_0^\infty t^2 g(t)\,dt.$$

Squaring the two sides of relation (1.5.12) and noticing that $X_n C_n = X_n$ and $C_n^2 = C_n$, we obtain:

$$X_{n+1}^2 = X_n^2 + B_n^2 + C_n + 2X_n B_n - 2X_n - 2B_n C_n. \qquad (1.5.13)$$

By considering the expectations of the two sides of (1.5.13), we obtain a relation giving $E[X_n]$:

$$0 = \rho + \lambda^2 \int_0^\infty t^2 g(t) dt + \rho + 2E[X_n]\rho - 2E[X_n] - 2\rho^2.$$

We obtain:

$$E[X_n] = \rho \left[1 + \frac{\rho(1 + Ks)}{2(1 - \rho)} \right] \tag{1.5.14}$$

where Ks is the square of the coefficient of variation of service time:

$$Ks = \mu^2 \int_0^\infty t^2 g(t) dt - 1.$$

1.6 - GI/M/1 QUEUE

The GI/M/1 queue can be considered as symmetrical with respect to the M/GI/1 queue. In fact the service time is now exponentially distributed and the arrival process is characterized by time intervals between two arrivals which are independent and identically distributed with density $f(t)$. We shall let $F(t) = \int_0^s f(s) ds$. The rate of service will always be denoted by μ and the mean time between two arrivals $\lambda^{-1} = \int_0^\infty t f(t) dt$. We shall specify this symmetry in the following.

Let X_n^* be the number of customers present in the system just before the instant a_n of the nth arrival.

Theorem

The chain $\{X_n^*\}$ is a Markov chain with transition matrix P^*. Let

$$p_k^* = \int_0^\infty \frac{e^{-\mu s}(\lambda s)^k}{k!} dF(s), \quad k = 0, 1, \ldots \tag{1.6.1}$$

then:

$$P^* = \begin{bmatrix} q_0^* & p_0^* & 0 & & & 0 \\ q_1^* & p_1^* & p_0^* & 0 & & . \\ q_2^* & p_2^* & p_1^* & p_0^* & 0 & 0 \\ . & . & . & . & & . \\ . & . & . & . & & . \\ . & . & . & . & & . \\ . & & & & & \end{bmatrix}$$

where $q_i^* = p_{i+1}^* + p_{i+2}^* + \cdots$.

Proof We must show that:

$$P\{X_{n+1}^* = i | X_0^*, X_1^*, \ldots, X_n^*\} = P\{X_{n+1}^* = i | X_n^*\}. \tag{1.6.2}$$

Let $D(t)$ be the number of departures from the queue in the interval $[0, t]$. We have:

$$X^*_{n+1} = X^*_n + 1 - [D(a_{n+1}) - D(a_n)]. \qquad (1.6.3)$$

$D(a_{n+1}) - D(a_n)$ represents the number of customers who have finished their service in the interval I_n. As the intervals between departures are exponentially distributed the quantity $D(a_{n+1}) - D(a_n)$ depends only on the number of customers at the instant a_n, that is X^*_n. The equality (1.6.3) verifies relation (1.6.2), which proves that $\{X^*_n\}$ is a Markov chain. The first column of the transition matrix is determined by

$$P\{X^*_{n+1} = 0 | X^*_n = i\} = P\{D(a_n + I_n) - D(a_n) = i + 1\}$$

$$= P\{D(I_n) = i + 1\} = \sum_{j=i+1}^{\infty} \int_0^\infty \frac{e^{-\mu s}(\lambda s)^j}{j!} f(s)\, ds = q^*_i.$$

$$= 0 \text{ otherwise.}$$

The kth column of the transition matrix is determined by:

$$P\{X^*_{n+1} = k | X^*_n = i\} = P\{D(a_n + I_n) - D(a_n) = i - k + 1\}$$

$$= P\{D(I_n) = i - k + 1\} = p^*_{i-k+1} \quad \text{if } i \geqslant k - 1, 0 \text{ otherwise.}$$

This concludes the proof of the theorem.

In the following we shall consider only aperiodic arrival processes in order to make the Markov chain $\{X^*_n\}$ aperiodic.

Let us now study the existence of a stationary solution.

Theorem

The chain $\{X^*\}$ is positive recurrent if and only if $\rho = \lambda/\mu < 1$.

The transition matrix is such that the Markov chain $\{X^*_n\}$ is irreducible and aperiodic. The theorem will be proved if the linear systems $\Pi^* P^* = \Pi^*$ and $\Pi^* 1 = 1$ have a unique strictly positive solution.

Let us write the two linear systems:

$$\begin{cases} \Pi^*_0 = p^*_1 \Pi^*_0 + p^*_2(\Pi^*_0 + \Pi^*_1) + p^*_3(\Pi^*_0 + \Pi^*_1 + \Pi^*_2) + \cdots, \\ \Pi^*_1 = p^*_0 \Pi^*_0 + p^*_1 \Pi^*_1 + p^*_2 \Pi^*_2 + \cdots, \\ \Pi^*_2 = p^*_0 \Pi^*_1 + p^*_1 \Pi^*_2 + p^*_2 \Pi^*_3 + \cdots, \end{cases} \qquad (1.6.4)$$

$$\ldots\ldots$$

and

$$\Pi^*_0 + \Pi^*_1 + \Pi^*_2 + \cdots = 1. \qquad (1.6.5)$$

Let us set:

$$b_k = \Pi^*_0 + \Pi^*_1 + \cdots + \Pi^*_{k-1}, \quad k = 1, 2, \ldots$$

By successively adding the terms of the equalities of system (1.6.4), we obtain:

$$\begin{cases} b_1 = p_1^* b_1^* + p_2^* b_2 + p_3^* b_3 + \cdots, \\ b_2 = p_0^* b_1 + p_1^* b_2 + p_1^* b_3 + \cdots, \\ b_3 = \qquad\quad p_0^* b_2 + p_1^* b_3 + \cdots, \\ \cdots \end{cases} \tag{1.6.6}$$

which can be written in condensed form:

$$b = \hat{P}*b \tag{1.6.7}$$

where $\hat{P}*$ is the matrix $P*$ with the first row and the first column transposed.

Notice that if λ is replaced by μ, μ by λ, and $f(x)$ by $g(x)$, $\hat{P}*$ becomes the matrix P from which the first row and the first column have been removed, which gives it the same asymptotic behaviour as P. Furthermore, to satisfy relation (1.6.5), it is established for $\lim_{n \to \infty} b_n = 1$. Relation (1.6.7) shows that b is the asymptotic state of a Markov chain with transition matrix P. Since $\lim_{n \to \infty} b_n = 1$, the chain is not ergodic which, from the study of the M/GI/1 queue, implies that $\rho \geqslant 1$. So that the linear system (1.6.6) has a unique solution with the condition $\lim_{n \to \infty} b_n = 1$, it is necessary and sufficient that the dual one has $\rho < 1$. This proves the theorem.

Before passing to the calculation of the mean number of customers and the system average response time, we shall compute explicitly the steady state probability described by $\Pi*$, for the embedded Markov chain.

Theorem

The steady state probability of the Markov chain $\Pi*$ is given by

$$\Pi_k^* = (1 - \sigma)\sigma^k, \quad k = 0, 1, 2, \ldots,$$

where σ is the unique solution between 0 and 1 of

$$\sigma = p_0^* + p_1^* \sigma + p_2^* \sigma^2 + \cdots = \mathscr{L}[f(\mu - \mu\sigma)].$$

Proof The solution of the linear system (1.6.6) is well known. It is of the form $b_i = 1 - \alpha^i, \alpha^i \in [0, 1]$. It only remains to determine α from one of the equations of the system. By taking the first we obtain:

$$\alpha = p_0^* + p_1^* \alpha + \cdots + p_k^* \alpha^k = \mathscr{L}[f(\mu - \mu\alpha)].$$

The function $h(\alpha) = p_0^* + p_1^* \alpha + \cdots + p_k^* \alpha^k$ is increasing and so is its derivative. Since $h(0) = p_0^* > 0$, $h(\alpha)$ intersects the function $k(\alpha) = \alpha$ at two points σ and 1. With point 1 corresponding to the case $\rho \geqslant 1$, there is only one solution to $h(\alpha) = k(\alpha)$ between 0 and 1, the solution of

$$\sigma = p_0^* + p_1^* \sigma + \cdots + p_k^* \sigma^k. \tag{1.6.8}$$

From this value we obtain:

$$\begin{cases} \Pi_0^* = b_1 = 1 - \sigma, \\ \Pi_1^* = b_2 - b_1 = (1 - \sigma)\sigma, \\ \Pi_k^* = b_{k+1} - b_k = (1 - \sigma)\sigma^k, \\ \dots\dots \end{cases} \qquad (1.6.9)$$

which is the required solution.

We must now study the stationary solution for any point. The method used is essentially the same as that used to study the M/GI/1 queue, based on the Markov renewal process.

Theorem

The process $\{X_n^*, a_n\}$ is a Markov renewal process with transition matrix $P^*(t)$:

$$P^*(t) = \begin{bmatrix} q_0^*(t) & p_0^*(t) & & (0) \\ q_1^*(t) & p_1^*(t) & p_0^*(t) \\ q_2^*(t) & p_2^*(t) & p_1^*(t) & p_0^*(t)\dots \\ \multicolumn{4}{c}{\dotfill} \end{bmatrix}$$

where

$$p_k^*(t) = \int_0^t \frac{e^{-\mu s}(\lambda s)^k}{k!}\, \mathrm{d}F(s), \quad k = 0, 1, \dots$$

and

$$q_k^*(t) = F(t) - \sum_{i=0}^{k} p_i^*(t).$$

Proof The reader will be able to establish the proof of this theorem by first proving the equality:

$$P\{X_{n+1}^* = j, a_{n+1} - a_n \leqslant t \mid X_0^*, \dots, X_n^*, a_0, \dots, a_n\}$$
$$= P\{X_{n+1}^* = j, a_{n+1} - a_n \leqslant t \mid X_n^*\}$$

for all integer n and j and t positive real,

then: $\qquad\qquad P\{X_{n+1}^* = 0, a_{n+1} - a_n \leqslant t \mid X_0^* = i\} = q_i^*(t),$

finally: $\qquad\qquad P\{X_{n+1}^* = k, a_{n+1} - a_n \leqslant t \mid X_0^* = i\} = p_{i-k+1}^*(t).$

We are again going to use the fundamental theorem of Markov renewal processes in order to obtain the stationary solution $p(i), i = 0, 1, \dots,$ at any instant.

We have:

$$p(k) = \lim_{t \to \infty} P\{N_t = k\} = \sum_j \frac{\Pi_j^*}{m^*} \int_0^{\infty} A^*(j, k, t)\, \mathrm{d}t. \qquad (1.6.10)$$

Π_j^* has been calculated previously, m^* is the mean time between two instants of the embedded Markov chain and $A^*(j, k, t)$ is the core of Markov renewal associated with $P^*(j, k, t)$.

If the embedded Markov chain is positive recurrent, that is if $\rho < 1$, the Markov renewal process is also positive recurrent and $m^* = \lambda^{-1}$.

$A^*(j,k,t)$, which is the probability that the system goes from state j to state k in the time interval $[0,t]$, can easily be calculated:

$$A^*(j,k,t) = \begin{cases} (1 - F(t))\dfrac{e^{-\mu t}(\mu t)^{j-k+1}}{(j-k+1)!} & \text{if } k > 0, j \geqslant k - 1, \\[2mm] (1 - F(t)) \displaystyle\sum_{n=j+1}^{\infty} \dfrac{e^{-\mu t}(\mu t)^n}{n!} & \text{if } j \geqslant k = 0, \\[2mm] 0 & \text{otherwise.} \end{cases} \qquad (1.6.11)$$

We can now prove the following theorem:

Theorem

If $\rho < 1$ then

$$\begin{cases} p(0) = 1 - \rho, \\ p(k) = \rho(1 - \sigma)\sigma^{k-1}. \end{cases}$$

Proof From the fundamental theorem stated earlier, we can write:

$$p(k) = \sum_j \lambda \sigma^j (1 - \sigma) \int_0^\infty A^*(j,k,t)\,\mathrm{d}t. \qquad (1.6.12)$$

Notice that:

$$\sigma = \sum_{j=0}^\infty p_j^* \sigma^j = \int_0^\infty e^{-\mu(1-\sigma)t} f(t)\,\mathrm{d}t = \mathscr{L}[f(\mu - \mu\sigma)].$$

For $k > 0$, we have:

$$p(k) = \lambda(1 - \sigma)\sigma^{k-1} \int_0^\infty \sum_j \sigma^{j-k+1} \frac{e^{-\mu t}(\mu t)^{j-k+1}}{(j-k+1)!}[1 - F(t)]\,\mathrm{d}t,$$

$$= \lambda(1 - \sigma)\sigma^{k-1} \int_0^\infty (1 - F(t))e^{-\mu(1-\sigma)t}\,\mathrm{d}t,$$

$$= \lambda(1 - \sigma)\sigma^{k-1}\left[\frac{1}{\mu(1-\sigma)}\left(1 - \int_0^\infty e^{-\mu(1-\sigma)t} f(t)\,\mathrm{d}t\right)\right].$$

From (1.6.13) we obtain the result for $k > 0$ and also for $k = 0$ by normalization.

In contrast to the M/GI/1 case, the stationary solution of the embedded Markov chain for the GI/M/1 queue is not identical to the stationary solution at any point. We have:

$$\rho\Pi_k^* = \sigma p(k) \qquad \text{if } k \neq 0$$

$$(1 - \rho)\Pi_0^* = (1 - \sigma)p(0).$$

The time distribution of the response of the GI/M/1 queue is obtained from the stationary solution of the embedded chain. Using Laplace transforms we have:

$$\mathscr{L}[T(x)] = \sum_{k=0}^{\infty} (1-\sigma)\sigma^k \left(\frac{\mu}{\mu+s}\right)^k = (1-\sigma)\mu \frac{\mu(1-\sigma)}{s+\mu(1-\sigma)},$$
$$T(x) = \mu(1-\sigma)e^{-\mu(1-\sigma)x}. \tag{1.6.14}$$

The distribution is exponential with coefficient $\mu(1-\sigma)$. The mean response time is:

$$E[T] = \frac{1}{\mu(1-\sigma)}; \tag{1.6.15}$$

by Little's formula we obtain:

$$E[N] = \frac{\rho}{1-\sigma}. \tag{1.6.16}$$

1.7 - GI/GI/1 QUEUE

This queue is much more complex and few explicit results are available in this case. Nevertheless we shall give a simple relation which exists between several of the usual parameters. Let x_n be the nth customer appearing in the system, I_n the time between the arrivals of x_n and x_{n+1}, S_n the service time of the nth customer and W_n the waiting time of the nth customer before his service starts.

In Figure 1.5 we present the timing diagram between the arrival of customer x_n and its exit from the system, assuming that customer x_{n+1} arrives between these two times. In Figure 1.6 we have shown the second possible case where

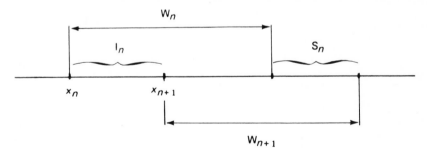

Figure 1.5 Case where customer x_{n+1} finds the server occupied

Figure 1.6 Case where customer x_{n+1} finds the server free

customer $n+1$ arrives at an empty system and is served immediately. In the latter case, it is evident that $W_{n+1}=0$. Returning to the first case, it appears from the diagram of Figure 1.6 that:

$$W_{n+1} = W_n + S_n - I_n. \tag{1.7.1}$$

We can combine the two solutions by:

$$W_{n+1} = \max(0, W_n + S_n - I_n) \tag{1.7.2}$$

hence $\{W_n, n = 0, 1, 2, \ldots\}$ is a Markov chain.

Relation (1.7.2) can be written in the form

$$W_{n+1} = \max(0, W_n + U_n) \quad \text{if} \quad U_n = S_n - I_n.$$

Developing this formula by recurrence and assuming that $W_0 = 0$, we obtain:

$$W_{n+1} = \max[0, U_n, U_n + U_{n-1}, U_n + U_{n-1} + U_{n-2}, \ldots, U_n$$
$$+ U_{n-1} + U_{n-2} + \cdots + U_0]. \tag{1.7.3}$$

Since the random variables S_n and I_n are independent, the $\{U_n\}$ form a sequence of independent and identically distributed random variables. Putting

$$V_j = U_n + U_{n-1} + \cdots + U_{n-j+1} \quad \text{and} \quad V_0 = 0.$$

We have

$$W_{n+1} = \max[V_0, V_1, \ldots, V_n]. \tag{1.7.4}$$

As n tends to infinity, W_{n+1} tends to the variable

$$W = \sup_i V_i$$

We have the following general theorem:

Theorem

Let $W(\sigma) = \lim_{n \to \infty} P\{W_n < \sigma\}$, the value of σ being bounded so that $W(\sigma)$ has a proper probability distribution if $\rho < 1$; $W(\sigma) = 0$ for all σ if $\rho > 1$ or if $\rho = 1$ and $U_n \neq 0$ with a positive probability.

More simply, the queue is stable if and only if $\rho < 1$ or $\rho = 1$ and $U_n = 0$. The latter case corresponds to equal and constant intervals of time between arrivals and service times.

By applying the appropriate law to the sequence V_i, with probability 1, we have:

$$\lim_{n \to \infty} \frac{V_n}{n} = E[U_i] \begin{cases} < 0 & \text{if } \rho < 1, \\ > 0 & \text{if } \rho > 1. \end{cases}$$

If $\rho > 1$, there exists a positive integer N such that $V_n \geq (1/2)nE[U_i]$ for all n. By making N sufficiently large, we have, with a probability of 1, $V_n \geq \sigma$ for all σ and hence $W(\sigma) = 0$.

Now if $\rho < 1$, then for all $x > 0$ and all $\varepsilon > 0$:

$$E[U_i] < 0;$$

which involves the existence of an integer N independent of x such that:

$$P\{V_n < x, n = N + 1, N + 2. ....\} \geqslant P\{V_n < 0, n = N + 1, N + 2, ...\} > 1 - (1/2)\varepsilon.$$

It is always possible to find an integer y such that for all $x > y$:

$$P\{V_n < x, n = 0, 1, ..., N\} > 1 - (1/2)\varepsilon.$$

It follows, by letting x tend to infinity, that $\lim W(x) = 1$. Since $W(x)$ is evidently a non-decreasing function and equal to zero for $x < 0$, this function is a proper probability distribution if $\rho < 1$.

This proves the principal part of the theorem. The proof that the queue is unstable if $\rho = 1$ and $U_n \neq 0$ with a positive probability is much more complicated to establish.

We shall prove this theorem again by using a theorem on the ergodicity of Markov chains and also its converse.

We are going to determine under which conditions the Markov chain $\{W_{n+1}\}$ determined by (1.7.2) is recurrent and positive. Let us assume that the arrival process for the queue is not periodic so that the Markov chain $\{W_n\}$ is irreducible and aperiodic.

Theorem

The Markov chain $\{W_n\}$ is positive recurrent if and only if $E[S_n] < E[I_n]$ (that is if $\rho < 1$).

Proof To prove the first part of this theorem, we will use a theorem concerning Markov chains giving a sufficient condition of ergodicity which is stated as follows (see Tweedie [10]).

A Markov chain W_n is ergodic if:

$$\gamma_w = E[|W_{n+1}| - |W_n| \,|\, W_n = w]$$

has an upper bound and if $\limsup_{w \to \infty} \gamma_w < 0$.

From relation (1.7.2), γ_w has an upper bound since by assumption $E[S_n]$ and $E[I_n]$ have. Also since w is sufficiently large we have:

$$\gamma_w = E[|W_{n+1}| - |W_n| \,|\, W_n = w] = E[S_n] - E[I_n].$$

From the theorem stated above, the Markov chain $\{W_n\}$ is ergodic if $E[S_n] < E[I_n]$.

Let us now show the converse: if $\{W_n\}$ is ergodic then $E[S_n] < E[I_n]$. If $\{W_n\}$ is ergodic we have:

$$P\left\{ \sum_{n=0}^{\infty} 1_{\{0\}}(W_n) = \infty \right\} = 1$$

that is the probability that the queue is empty is strictly positive. Since $p(0) = 1 - \rho$ (cf. 1.3), this implies

$$E[S_n] < E[I_n]$$

which proves the theorem.

In the case of a periodic chain, obtaining the necessary and sufficient condition depends on the initial conditions. If the number of customers in the system at the initial state is finite, the necessary and sufficient condition is: $\lambda \leqslant \mu$.

To conclude this part on the general GI/GI/1 queue, we give an approximate formula for the mean number of customers in the system:

$$E[N] \approx \rho \left(1 + \frac{\rho(Ka + Ks)}{2(1 - \rho)} \right),$$

where

$$Ka = [E[I^2] - (E[I])^2/(E[I])^2,$$
$$Ks = (E[S^2] - (E[S])^2)/(E[S])^2.$$

This formula was originally established by Kingman. The proof uses relation (1.7.2):

$$W_{n+1} = \max(0, W_n + S_n - I_n).$$

Let us put $X^+ = \max(0, X)$ and $X^- = -\min(0, X)$. We again take $U_n = S_n - I_n$. Assuming that $W = \sup_{i \geqslant 0}(V_i)$ and $U = \lim U_n$ exist and have finite values, we can write from relation (1.7.2):

$$E[W] = E[(W + U)^+] \quad \text{and} \quad \sigma_W^2 = \sigma_{(W+U)^+}^2$$

(σ_W being the standard deviation of the random variable W).

From this last equality, one obtains:

$$\sigma_W^2 + \sigma_U^2 = \sigma_{(W+U)^+}^2 + \sigma_{(W+U)^-}^2 + 2E((W+U)^-)E((W+U)^-),$$
$$= \sigma_W^2 + \sigma_{(W+U)^-}^2 + 2E(-U)E(W),$$

or again:

$$\sigma_U^2 = \sigma_{(W+U)^-}^2 - 2E(U)E(W)$$

which we can put in the form:

$$E[W] = \frac{\sigma_U^2}{2\lambda(1 - \rho)} - \frac{\sigma_{(W+U)^-}^2}{2\lambda(1 - \rho)}, \qquad (1.7.5)$$

by noticing that

$$E[U] = -\lambda(1 - \rho).$$

Assuming the second term to be negligible we obtain, returning to conventional notation:

$$E[W] \approx \frac{\rho(Ka + Ks)}{2\mu(1 - \rho)}.$$

Including the service time and using Little's formula, we obtain the mean number of customers in the queue:

$$E[N] = \rho \left[1 + \frac{\rho(Ka + Ks)}{2(1 - \rho)} \right].$$

It should be noted that this formula provides:

1. An upper bound since the term which we have neglected is positive.
2. The Pollaczek–Khintchine formula for the M/GI/1 case since $Ka = 1$.

1.8 - APPLICATION TO THE EVALUATION OF THE CHARACTERISTICS OF SOME COMPUTER SYSTEMS

The theory of queues is particularly well adapted to the study of the performance of computer systems. In fact, the performance of such systems depends on the hardware resources (central processing units, memory, etc.), on software (system programs, compilers, etc.) and on the organization and management of these resources. In view of the increasing complexity of computing systems, it is more and more difficult to predict their behaviour and performance without making use of simulation or mathematical models. In such models, it is convenient to represent the resources as 'servers' and the programs as 'customers'.

In the following sections we provide several simple examples of the application of queueing theory to the performance evaluation of computer systems.

1.8.1. Response time of a computer system

Consider a single program computer using the central processing unit as its only resource. As shown in Figure 1.7, n terminals are connected to this computer. The terminals can be geographically separated and can be either visual display units or printing terminals.

Assume that the processing time of a task on the central processing unit is exponentially distributed with a mean of 500 milliseconds (ms). The time interval

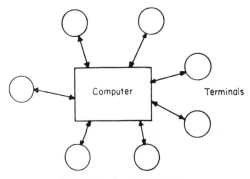

Figure 1.7 Computer system

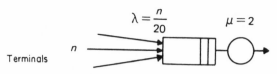

Figure 1.8 Mathematical model of the computer system represented in Figure 1.7

between two requests from the same terminal is also assumed to follow an exponential distribution with a mean of 20 seconds. The mathematical model which can be associated with this system is shown in Figure 1.8. It is an M/M/1 queue with a service rate of 2 per second and an arrival rate of $\frac{1}{20} \times n$ since the superposition of independent Poisson streams is Poissonian.

In Figure 1.9 we have shown the response time for a task as a function of the number of terminals n.

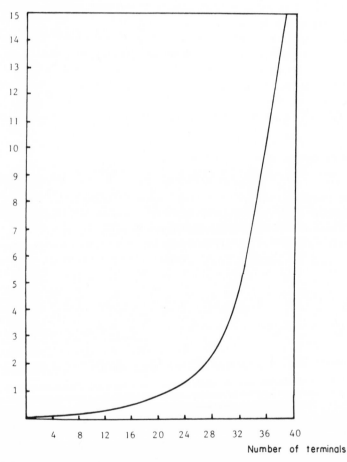

Figure 1.9 Response time for a task as a function of the number of terminals

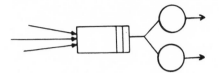

Figure 1.10 Mathematical model of a dual processor computer

This response time is simply:

$$E[T] = \frac{1}{\mu(1-\rho)} \quad \text{where} \quad \rho = \frac{\lambda}{\mu} = \frac{n}{40}.$$

For the system to be able to carry out all tasks, it is necessary that $\rho < 1$, that is the number of terminals must be strictly less than 40.

For instance, assume that the number of terminals is equal to 36. The response time $E[T]$ is then 5 seconds.

Assume that the computer, instead of having a single processor, has two processors, each being identical to that described previously. The mathematical model becomes that shown in Figure 1.10. It is an M/M/2 queue.

In this case the mean response time, if there are 36 terminals, becomes:

$$E[T] = 0,28 \text{ s}$$

(see the exercises at the end of the chapter for the result).

1.8.2. Response time for a disc access

Assume that a disc receives a mean of k read demands per second with arrivals following a Poisson process. After processing by the central processing unit of the computer, a read operation generates a write demand in 30% of cases and two write demands in other cases.

We shall assume that the read and write times are equal with a mean of 1 millisecond. The mean time to position the read head on the drum is 50 milliseconds and the disc makes one complete revolution in 12 milliseconds; we propose to find the mean service delay, assuming that the distribution of access times to the data is an Erlang-m distribution. The mean service time for a disc access amounts to:

$$E[S] = \text{positioning of the arm} + 1/2 \text{ revolution of the disc}$$
$$+ \text{ read or write time,}$$
$$= 57 \text{ milliseconds.}$$

So the rate of service per second is $1/0.057 = 17.5 = \mu$. The mean number of disc accesses per second is:

$$k \times (1 \times 0.3 + 2 \times 0.7) = 1.7k = \lambda$$

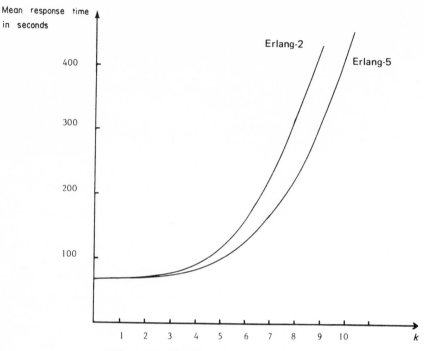

Figure 1.11 Response time of the disc

and the traffic rate is then:

$$\rho = \lambda/\mu = 0{,}097k:$$

Using the Pollaczek–Khintchine formula, one has:

$$E[T] = \frac{1}{\mu}\left[1 + \frac{\rho(1 + Ks)}{2(1 - \rho)}\right].$$

In Figure 1.11 we give the response time as a function of the number of read operations per second, k, for two values of the parameter m of Erlang's distribution.

1.8.3. Model of a 'send and wait' protocol

In this section we give a much more detailed example of the evaluation of the performance of a certain arrangement used in transmission networks: a protocol for communication between two nodes in the network.

There are several techniques for transmitting information from one point to another. One frequently used method in computer networks is packet switching. The data to be transmitted are separated into packets having a fixed or variable length; but the maximum length is permanently fixed for each network (generally 256 bytes). These packets are carried by a *packet-switching network* which allows

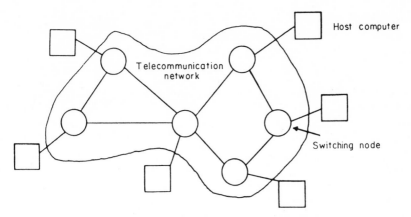

Figure 1.12 Example of a data transmission network

interconnection of all the computers on the network by way of an array of switching nodes. We represent such a computer network in Figure 1.12.

The packets travel from switching node to switching node in order to reach their destination. A liaison protocol must ensure that packets leaving one node arrive correctly at the following node.

Before being able to model the behaviour of a liaison protocol between two nodes, it is necessary to describe the operation of a node suitable for packet switching. Figure 1.13 shows the simplified general organization of the logical

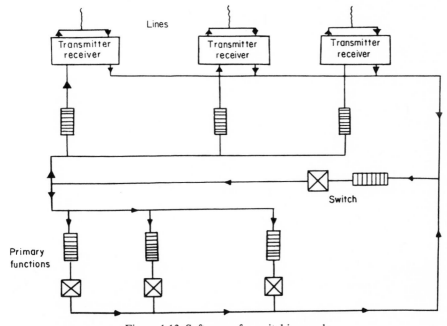

Figure 1.13 Software of a switching node

structure of a node. It is the one used in the packet communication network (called CIGALE) of the computer network CYCLADES built in France in the seventies.

The software functions of a node can be summarized in the following way:

–a transmitter is associated with each transmission line. This manages the output line and sends packets on the line following the adopted procedure. The receiver must receive the packets from the line observing the transmission procedure, verify that the packet is correct, acquire empty buffers and finally transfer the packets to the switching queue;

–the functions of the switch are as follows: to manage its queue (which is unique), to determine, as a function of the destination of the packet, the best output line, to transfer the packet to the queue for this output line, to destroy the packet if this queue is full;

–the primary functions represent a set of services which can be carried out by the node. In CIGALE for example there are 10 functions such as:

●ECHO   which sends the received packets back to their origin,
●STAT   which sends statistical data,
●HEURE which manages the time in CIGALE (HOUR in English),
●ROUTE which manages the adaptive routing,
etc.

The various functions of the node software are subject to priorities. Different priority levels are associated with them corresponding to control and monitoring. The principal priorities are, in increasing order:

1. primary functions,
2. switching,
3. line transmission.

The 'send and wait' protocol is described as follows. In the following we shall call it the SW protocol, the most commonly used notation. Its essential characteristic consists of sending a new packet only when the preceding packet has clearly been acknowledged. In the case of a transmission error, a copy of the erroneous packet is retransmitted before anything new is sent.

Figure 1.14 gives a diagram of the mode of operation of the SW procedure as a function of time. When the packet reaches the receiver, after a suitable management time ('overhead'), an acknowledgement must be sent to the node transmitter. A special acknowledgement 'frame' is then constructed. However, it is necessary to wait for the end of transmission of the previous packet if one is being sent. This time delay is denoted by τ in the diagram of Figure 1.14. The length of this frame is generally 7 bytes.

The acknowledgement does not pass through the switch; it is directly processed by the transmitter. The quantities ω_{01} and ω_{11} are the management times of the transmitter and receiver, while ω_{00} and ω_{10} represent the management time necessary for decoding, monitoring, etc., on arrival at a node.

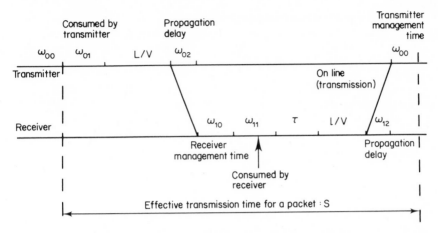

Figure 1.14 Mode of operation of the SW procedure

We denote the service time for one transmission by S. This time can be written in the following form:

$$S = \omega_{00} + \omega_{01} + \omega_{02} + L/V + \omega_{10} + \omega_{11} + \omega_{12} + \ell/V + \tau,$$

where L is the total length of the packet,
ℓ is the length of the packet which sends the acknowledgement,
V is the capacity of the line.

We shall use values obtained from measurements on the CYCLADES network:

$$\omega_{00} = \omega_{10} \approx 5 \text{ milliseconds}$$
$$\omega_{01} = \omega_{11} \approx 3 \text{ milliseconds}$$

For the propagation times $(\omega_{02}, \omega_{12})$ we shall use 10 milliseconds for a line with a binary rate of 4.8 kbits/second and 3 milliseconds for other lines. These times take account of the management times of the modems used.

These propagation times correspond to lines with a length of approximately 500 km. If a transmission error occurs, the packet is destroyed and a copy, stored in the transmitting node, is retransmitted. Let us denote the probability of a transmission error by q.

It is important to notice that the effective time S varies as a function of the values of ℓ and τ. We shall take ℓ equal to 7 bytes. The quantity τ takes a value of half a packet length if the line is very busy.

To give an idea of the values of transmission times, we shall assume that the mean length of a packet is 1000 bits. The mean service time is given in Table 1.1 as a function of the capacity of the line.

If the probability q is not zero, it is necessary to retransmit the erroneous packets. Assuming that the error is detected immediately, and a new transmission can start without delay, the time actually required for transmission of a packet

Table 1.1 - **Mean transmission times**

Line capacity	4.8 kbits/s	19.2 kbits/s	48 kbits/s
$\bar{S}$	356 ms	103 ms	54 ms

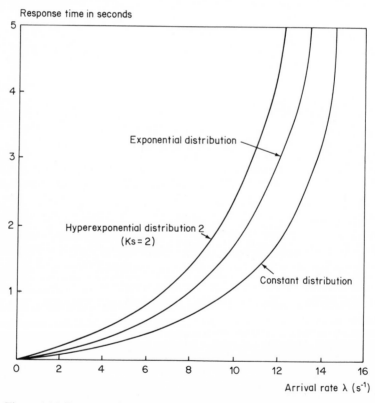

Figure 1.15 Model of the SW procedure

has a mean value of $\bar{S}/1 - q$. In the following we shall set $\mu = (1 - q)/\bar{S}$. We are now able to construct a simple mathematical model representing the operation of the SW procedure. The packets arrive in a queue with a rate λ and are served at a rate μ. The model is represented by Figure 1.15.

We assume that the arrival of packets at a switching node follows a Poisson process. In contrast the service process depends very much on the length of the

Figure 1.16 Response time of a switch using the SW procedure $q = 0.1$

packets to be transmitted. In fact for a constant line transmission speed, the transmission time for a packet is a function of its length. The distribution of service times depends on the type of packets used by the network. If the network is transmitting time shared messsages, the packets are small and the service distribution is certainly sufficiently random to be close to exponential. If file transfers are made, all packets will have the maximum length of 256 bytes, and the distribution of service times is well approximated by a constant law. Finally, if several kinds of traffic are superposed, the distribution will have an increased variance and will approach a hyperexponential one. In summary, according to the type of traffic, the square of the variation coefficient of the service time distribution can vary from 0 to 2 or 3. In Figure 1.16 we have plotted the curve for the response time of a switching node using the SW procedure, for different values of arrival rate and mean service time.

EXERCISES

1. Consider an M/M/C queue: the arrivals are Poisson, the services are exponential and the number of service units is equal to C. Find the probability of the length of the queue in the stationary state and the mean number of customers in the system.

Answer

Let

$$u = \frac{\lambda}{\mu} \quad \text{et} \quad \rho = \frac{\lambda}{C\mu}$$

then:

$$p(0) = \left[\sum_{k=0}^{C-1} \frac{u^k}{k!} + \frac{u^C}{C!(1-\rho)} \right]^{-1},$$

$$p(k) = \begin{cases} \dfrac{u^k}{k!} p(0) & \text{if } k = 0, 1, \ldots, C, \\[3mm] \dfrac{u^k}{C! C^{k-C}} p(0) & \text{if } k \geqslant C, \end{cases}$$

$$E[N] = \frac{\rho u^C}{C!(1-\rho)^2} p(0) + u.$$

2. Consider an M/M/1/K queue, a Markovian queue which cannot contain more than K customers. When the queue is full, a new customer that arrives is not allowed to join the queue and is rejected. Calculate the probabilities in the stationary state of the number of customers in the queue and also the mean number of customers in the system.

Answer

$$p(k) = \frac{\rho^k(1-\rho)}{1-\rho^{K+1}} \quad \text{or} \quad \rho = \frac{\lambda}{\mu},$$

$$E[N] = \frac{\rho}{1-\rho} - \frac{(K+1)\rho^{K+1}}{1-\rho^{K+1}}.$$

3. Study the M/M/C/C queue.

Answer

$$p_n = \frac{\rho^n/n!}{1+\rho+\rho^2/2!+\cdots+\rho^C/C!}, \quad \rho = \frac{\lambda}{\mu}.$$

Comment p_C gives the probability that a new customer is rejected; that is Erlang's formula known to telephone engineers.

4. Study the M/M/∞ queue.

Answer

$$p_n = e^{-\rho}(\rho^n/n!).$$

5. Show that for the M/M/1 queue, the probability that there are n or more customers in the system is ρ^n, and the mean number of customers in the system knowing that there is at least one customer being served is given by $1/(1-\rho)$.

6. Give another proof for Little's formula for an M/G/1 queue by using the properties of the Poisson process.

7. Study the M/D/1 queue (constant or deterministic service).

Answer

$$p_0 = 1-\rho.$$
$$p_1 = (1-\rho)(e^\rho - 1),$$

for $n \geqslant 2$,

$$p_n = (1-\rho)\sum_{k=1}^{n}(-1)^{n-k}e^{k\rho}\left[\frac{(k\rho)^{n-k}}{(n-k)!} + \frac{(k\rho)^{n-k-1}}{(n-k-1)!}\right].$$

8. Calculate the mean length of a period during which there are no customers in the system and of a period during which there is always at least one customer being served, for the M/G/1 queue.

9. Study the M/E$_2$/1 queue where the Erlang-2 server is separated into one set of two stations with an exponentially distributed service time of mean $(2\mu)^{-1}$ as shown in the following figure:

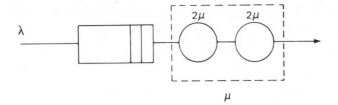

In the array of two service stations, there cannot be more than a single customer.

10. Consider an M/M/1 queue which can contain two classes of customer C1 and C2. The class C1 customers have an absolute priority over class C2 customers, that is if a class C2 customer is being served when a class C1 customer arrives at the station, the class C1 customer takes the place of the class C2 customer who goes back to the head of the queue and will resume service when the C1 customer has left. Calculate the mean number of class C1 and C2 customers present in the system. The proportion of class C1 customers in the arrivals is α and the mean service times are μ_1 and μ_2.

Answer

$$\rho_1 = \frac{\alpha\lambda}{\mu}, \quad E(N_1) = \frac{\rho_1}{1 - \rho_1};$$

$$E(N_2) = \frac{\rho_2}{1 - \rho_1 - \rho_2}\left[1 + \frac{\rho_1}{1 - \rho_1}\frac{\mu_2}{\mu_1}\right],$$

$$\rho_2 = \frac{(1 - \alpha)\lambda}{\mu}.$$

BIBLIOGRAPHY

The first work on a simple queue with a single server dates from the beginning of the century with Erlang. However, it was necessary to wait another 30 years for the first general results on specific queues [1] and [2]. Most of the results described in this chapter have been obtained in the fifties in more or less complicated form; see for instance [3] to [8].

The formulations presented in the text for M/G/1 and G/M/1 queues arise principally from the works of Cinlar [9] and for the G/G/1 queue in part from those of Tweedie [10].

The first part of this chapter is a new presentation which unifies, in a certain sense, the works on operational analysis [11].

1. Pollaczek, F. (1930). Über eine Aufgabe der Wahrscheinlichkeitstheorie I, II, *Math. Zeitschrift.*, **32**, 64–100, 729–750.
2. Khinchin, A. Y. (1932). Mathematical theory of stationary queues, *Math. Sbornik*, **39**, 73–84.
3. Kendall, D. G. (1951). Some problems in the theory of queues, *Journal of the Royal Statistical Society Ser. B.* **13**, 151–185.

44

4. Lindley, D. V. (1952). The theory of queues with a single server, *Proc. Cambridge Philosophical Society*, **48**, 277–289.
5. Takacs, L. (1955). Investigation of waiting time problems by reduction to Markov processes, *Acta Math. Acad. Sci. Hung*, **6**, 101–129.
6. Benes, V. E. (1956). On queues with Poisson arrivals, *Annals Math. Stat.*, **28**, 670–677.
7. Gaver, D. P. (1959). Imbedded Markov chain analysis of a waiting line process in continuous time, *Annals Mathematical Statistics*, **30**, 698–720.
8. Little, J. D. C. (1961). A proof of the queueing formula $L = \lambda W$, *Operations Research*, **9**, 383–387.
9. Cinlar, E. (1975). Introduction to Stochastic Processes, Englewoods Cliffs, N. J. Prentice Hall.
10. Tweedie, R. L. (1975). Sufficient conditions for ergodicity and recurrence of Markov chain on a general state space, *Stoch. Proc. and their Appl.*, **3**, 385–403.
11. Denning, P. J., and Buzen, J. P. (1978). The operational analysis of queueing network models, *Computing Surveys*, **10**, 225–261.

CHAPTER 2

Jackson Networks

2.1 - INTRODUCTION

This chapter is entirely devoted to a study of the 'simplest' queueing networks. In the random case, for we treat deterministic and stochastic cases, it is concerned with systems consisting of N service stations and a queue of unbounded length at each station. The advance of a customer from one station to another is represented by a Markov chain and no distinction is made between the stochastic characteristics of the customers. That is, if the series of stations visited by any customer is denoted by $(Y_i)_{i \geqslant 1}$, the conditional probability law for Y_i has the following property:

$$P\{Y_{i+1}|Y_1, Y_2, \ldots, Y_i\} = P\{Y_{i+1}|Y_i\}$$

for any $i \geqslant 1$. Of course, each Y_i is the number of a service station, that is $Y_i \in \{1, 2, \ldots, N\}$, or $Y_i = N + 1$ to represent departure to the 'exterior' of the network. Thus we admit open systems (which receive customers from the exterior according to a Poisson process and send them to the exterior) and closed systems which have a constant number of customers. In the stochastic case, the service times for each queue will always be independent of each other and distributed according to exponential laws having parameters which can depend on the length of the respective queue.

This model, known as a Jackson network, has the interesting property of having a stationary solution in 'product form' for the joint probabilities of the lengths of the queues: it is written in the form of a product of the marginal distributions of each queue. This is why it is very often used to model very complex computer systems and data transmission systems. It is interesting to note that it took fifty years from the first appearance of queueing theory for this simple and surprising result to be discovered.

However, we start in Section 2.2 with a *deterministic* system observed during a defined time interval. From assumptions concerning transitions between states for this model, we obtain the properties of the proportions of time spent

45

in the various states (these proportions of time replacing the probabilities in a stochastic model), which are formally and numerically identical to the 'product form' of Jackson networks.

In the case of a closed system, the resulting form does not lend itself to direct numerical calculation: in fact it is necessary to be able to calculate the sums of an 'exponential' number of terms. This is why, having examined a particular closed system in Section 2.3, Section 2.5 is devoted to the development of effective algorithms in time and space for calculation of interesting quantities related to the model. The algorithms are identical for stochastic and deterministic models.

Finally, in Section 2.6, stochastic results are described and demonstrated, and several application examples are developed in detail.

2.2 - DETERMINISTIC ANALYSIS OF AN OPEN QUEUEING NETWORK

We consider a sufficiently general model which allows us to represent the behaviour of a wide range of computer systems and data transmission systems, and to evaluate their characteristics.

It consists of a queueing network containing N stations. We denote the number of customers at station i by k_i the state vector of the system being:

$$k = (k_1, \ldots, k_N).$$

Changes of state are caused by arrivals of customers from the 'exterior', departures to the exterior and ends of service followed by departures for another queue.

Following an approach already adopted in Chapter 1, we examine the system during a given time interval of length T such that the states of the system at the beginning and end of the interval are the same.

The number of transitions between any two states k and k' during this interval will be denoted by $D(k, k')$. For any state k, we shall have:

$$\sum_{k'} D(k', k) = \sum_{k'} D(k, k').$$

That is, the number of transitions (*) into state k will be equal to the number of transitions from the same state.

Let $T(k)$ be the time spent in state k during the interval and $d(k, k')$ be the transition rate from k to k':

$$d(k, k') = \frac{D(k, k')}{T(k)}.$$

---

*Here, and in the following, a transition is the passage of the system, *in a single jump*, from one state to a neighbouring (or accessible) state.

Therefore we have:

$$\sum_{k'} T(k')d(k',k) = \sum_{k'} T(k)d(k,k')$$

or:

$$\sum_{k'} p(k')d(k',k) = \sum_{k'} p(k)d(k,k') \qquad (2.2.1)$$

if the fraction of time spent in state k is defined by:

$$p(k) = \frac{T(k)}{T}.$$

Equation (2.2.1) establishes the global equilibrium of the queueing network during the interval. We shall now make several restrictive assumptions in respect of $d(k,k')$:

(i) Assumption of limited transitions

$$d(k',k) = 0$$

except: if, for any i, k' is equal to any of the following states:

(a) $a(k,i) = (k_1,\ldots,k_i+1,\ldots,k_N)$ and $k = (k_1,\ldots,k_i,\ldots,k_N)$

or

(b) $b(k,i) = (k_1,\ldots,k_i-1,\ldots,k_N)$ and $k = (k_1,\ldots,k_i,\ldots,k_N)$

or, if for any $j \neq i$:

(c) $c(k,i,j) = (k_1,\ldots,k_i+1,\ldots,k_j-1,\ldots,k_N)$ and
$$k = (k_1,\ldots,k_i,\ldots,k_j,\ldots,k_N).$$

In cases (a), (b) and (c), $d(k,k')$ can be positive or zero.

This assumption specifies quite simply that changes of state arise from the passage of *one* customer at a time from one queue to another, or by *a departure* to the exterior or *an arrival* from the exterior.

(ii) Assumption of local dependence

For the k and k' cited in (i) for which one can have $d(k',k) > 0$, we have

$$d(k',k) = \begin{cases} p_{i,N+1}X_i(k_i+1), & \text{if } k' = (k_1,\ldots,k_i+1,\ldots,k_N), \quad \text{(a)} \\ p_{0,i}X_0, & \text{if } k' = (k_1,\ldots,k_i-1,\ldots,k_N), \quad \text{(b)} \\ p_{ij}X_i(k_i+1) & \text{if } k' = (k_1,\ldots,k_i+1,\ldots, \\ & \qquad\qquad k_j-1,\ldots,k_N), \quad \text{(c)} \end{cases}$$

where $k = (k_1,\ldots,k_i,\ldots,k_j,\ldots,k_N)$, $i \neq j$, the $p_{0,i}$ and the $p_{i,j}$ are real and non-negative such that:

$$\sum_{i=1}^{N} p_{0,i} = 1$$

and for all i, we have:

$$p_{ii} = 0 \quad \text{and} \quad \sum_{j=1}^{N+1} p_{ij} = 1.$$

Finally the $X_i(\ell)$ are real and non-negative such that:

$$X_i(\ell) = 0 \quad \text{if} \quad \ell = 0.$$

This assumption implies, that the transition rate $d(k, k')$ can depend only on i, j and the number of customers $(k_i + 1)$ in the queue i at the instant preceding the departure.

With these two assumptions, we shall show that equation (2.2.1) has the following solution:

$$p(k) = G^{-1} \prod_{i=1}^{N} \prod_{m=1}^{k_i} \left[\frac{e_i X_0}{X_i(m)} \right] \tag{2.2.2}$$

where G is a normalization constant which makes:

$$\sum_k p(k) = \frac{\sum_k T(k)}{T} = \frac{T}{T} = 1, \tag{2.2.3}$$

that is:

$$G = \sum_k \prod_{i=1}^{N} \prod_{m=1}^{k_i} \left[\frac{e_i X_0}{X_i(m)} \right]$$

where $k = (k_1, k_2, \ldots, k_N)$, $k_i \geq 0$, and where the e_i are solutions to the system of equations:

$$e_i = p_{0,i} + \sum_{j=1}^{N} e_j p_{ji}, \quad i = 1, \ldots, N. \tag{2.2.4}$$

To prove formula (2.2.2), which allows us to calculate the proportion of time $p(k')$ spent in any state k' in the interval of length T, let us consider (2.2.1). It consists of a system of linear equations. To prove (2.2.2), it is sufficient to show that (2.2.2) is a solution of (2.2.1); in fact it is the only solution which satisfies equality (2.2.3). The only pairs k, k' to be considered are those which satisfy assumption (i), and in this case the relation $p(k')/p(k)$ can be written by using (2.2.2):

$$\frac{p(k')}{p(k)} = \begin{cases} e_i X_0 / X_i(k_i + 1), & \text{(a)} \\ X_i(k_i)/e_i X_0, & \text{(b)} \\ e_j X_j(k_j)/X_i(k_i + 1)e_j & \text{(c)} \end{cases} \tag{2.2.5}$$

corresponding, respectively, to the cases (a), (b) and (c) of assumption (i). By introducing (2.2.5) into (2.2.1) with assumption (ii), one obtains:

$$\sum_{i=1}^{N} \frac{e_i X_0}{X_i(k_i+1)} p_{i,N+1} X_i(k_i+1) + \sum_{i=1}^{N} \frac{X_i(k_i)}{e_i X_0} p_{0,i} X_0$$

$$+ \sum_{\substack{i,j=1 \\ i \neq j}}^{N} \frac{e_i X_j(k_j)}{e_j X_i(k_i+1)} p_{ij} X_i(k_i+1) = \sum_{i=1}^{N} p_{i,N+1} X_i(k_i)$$

$$+ \sum_{i=1}^{N} p_{0,i} X_0 + \sum_{\substack{i,j=1 \\ i \neq j}}^{N} p_{ij} X_i(k_i).$$

After simplification this becomes:

$$\sum_{i=1}^{N} \left[X_0 e_i p_{i,N+1} + \frac{p_{0,i} X_i(k_i)}{e_i} + \sum_{\substack{j=1 \\ j \neq i}}^{N} \frac{e_i}{e_j} p_{ij} X_j(k_j) \right]$$

$$= \sum_{i=1}^{N} \left[X_i(k_i) p_{i,N+1} + p_{0,i} X_0 + \sum_{\substack{j=1 \\ j \neq i}}^{N} p_{ij} X_i(k_i) \right],$$

$$= \sum_{i=1}^{N} [p_{0,i} X_0 + X_i(k_i)].$$

By using (2.2.4), one obtains:

$$\sum_{i=1}^{N} \left[X_0 e_i p_{i,N+1} + \frac{p_{0,i} X_i(k_i)}{e_i} + \sum_{j=1}^{N} X_j(k_j) \left(\frac{e_j - p_{0,j}}{e_j} \right) \right.$$

$$\left. = \sum_{i=1}^{N} [p_{0,i} X_0 + X_i(k_i)] \right].$$

This equality is satisfied if:

$$\sum_{i=1}^{N} X_0 e_i p_{i,N+1} = \sum_{i=1}^{N} p_{0,i} X_0$$

that is if:

$$\sum_{i=1}^{N} e_i p_{i,N+1} = 1.$$

But this can also be written:

$$\sum_{i=1}^{N} \left(e_i - e_i \sum_{j=1}^{N} p_{ij} \right) = 1$$

or

$$1 = \sum_{i=1}^{N} e_i - \sum_{j=1}^{N} \sum_{i=1}^{N} e_i p_{ij} = \sum_{i=1}^{N} e_i - \sum_{j=1}^{N} (e_j - p_{0,j}) = 1,$$

which completes the proof of (2.2.2).

2.3 - THE CENTRAL SERVER SYSTEM: EXAMPLE
OF A CLOSED SYSTEM

A particular case of the model studied in the preceding paragraph is very useful for modelling operational systems. It consists of a system with central server presented schematically in Figure 2.1.

Station 1 represents the central processing unit of a computer operating in multiprogramming mode. Stations 2 to N represent the peripheral units (discs, input–output units, etc.). The system is 'closed', that is a fixed number of customers K move about the system such that:

$$\sum_{i=1}^{N} k_i = K,$$

for all states $k = (k_1, \ldots, k_N)$. From the model studied in Section 2.2, we see that arrivals from the exterior of the system and departures to the exterior are suppressed. On the other hand, to represent transitions between system queues, the probabilities p_{ij} take a particular form imposed by the topography of Figure 2.1:

$$p_{ij} = \begin{cases} 1 & \text{if } j = 1, \quad i \neq 1, \\ p_j & \text{if } i = 1, \quad j \neq 1, \\ 0 & \text{for all other cases,} \end{cases}$$

and hence $\sum_{j=1}^{N} p_{ij} = 1$ for all $i = 1, \ldots, N$.

The equilibrium equation (2.2.1) for the system is unchanged: for each k,

$$\sum_{k'} p(k') \mathrm{d}(k', k) = \sum_{k'} p(k) \mathrm{d}(k, k'),$$

but the transition rates $\mathrm{d}(k', k) = D(k', k)/T(k')$ will necessarily have a particular

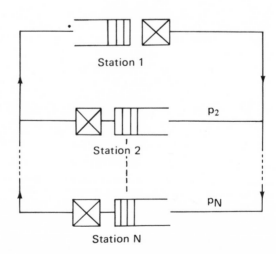

Figure 2.1 N station central server model

form, taking account of the fact that the system is closed (the total number of customers in the system remains constant) and the specific nature of the p_{ij}. For $j = 2,\ldots,N$, they can be written:

$$d(k',k) = \begin{cases} p_j X_1(k_1 + 1) & \text{if} \quad k' = (k_1 + 1,\ldots,k_j - 1,\ldots,k_N), \\ X_j(k_j + 1) & \text{if} \quad k' = (k_1 - 1,\ldots,k_j + 1,\ldots,k_N). \end{cases} \tag{2.3.1}$$

We shall show that the solution of the equilibrium equation takes the following form:

$$p(k) = G(K,N)^{-1} \prod_{i=1}^{N} \prod_{m=1}^{k_i} \left[\frac{p_i}{X_i(m)} \right] \tag{2.3.2}$$

where to simplify the type we have replaced 1 by p_1. Hence:

$$G(K,N) = \sum_k \prod_{i=1}^{N} \prod_{m=1}^{k_i} \left[\frac{p_i}{X_i(m)} \right]$$

and the normalization constant allows that:

$$\sum_k p(k) = 1$$

where each $k = (k_1,\ldots,k_N)$ obeys:

$$k_i \geqslant 0, \quad \sum_{i=1}^{N} k_i = K.$$

To prove (2.3.2), we use the equilibrium equation in the following form:

$$\sum_{k'} \frac{p(k')}{p(k)} d(k',k) = \sum_{k'} d(k,k')$$

and we verify that (2.3.2) is the solution to it. Taking account of the values taken by the $d(k',k)$ in (2.3.1), this can be written:

$$\sum_{j=2}^{N} \left(\frac{X_j(k_j)}{p_j} \cdot \frac{1}{X_1(k_1 + 1)} \right) p_j X_1(k_1 + 1)$$

$$+ \sum_{j=2}^{N} \left(\frac{p_j}{X_j(k_j + 1)} \cdot X_1(k_1) \right) X_j(k_j + 1)$$

$$= \sum_{j=2}^{N} [p_j X_1(k_1) + X_j(k_j)].$$

Since this equality is evidently satisfied, we conclude that (2.3.2) is indeed the unique solution of the equilibrium equations.

A parameter which often interests the user is the *rate of utilization* of the central processing unit or of any other peripheral unit; it is the proportion of time during which the unit will be active (or will contain at least one customer) compared with the total time T. The rate of utilization of station j is therefore defined as:

$$U_j = \sum_{\substack{k \\ k_j > 0}} T(k)/T = \sum_{\substack{k \\ k_j > 0}} p(k), \tag{2.3.3}$$

for $j = 1,\dots,N$. By using (2.3.2), we have:

$$U_j = \frac{p_j G(K-1,N)}{X_j G(K,N)}, \tag{2.3.4}$$

in the case where $X_j(k_j) \equiv X_j$ that is where X_j does not depend on k_j. This example shows the importance of the normalization constant $G(K,N)$ which will be calculated in Section 2.5.

2.4 - SOLUTION OF A GENERAL CLOSED SYSTEM

In the preceding section, we have examined the behaviour of a particular closed system containing N stations or queues and a constant number of customers K. The model was restrictive due to the fact that customers could not move from a queue $j = 2,\dots,N$ except to queue 1. If one allows customers to transfer between any queues, one obtains the model which we shall study in this section. It is characterized, over a time interval of length T such that the state is the same at the beginning and end of the interval, by the equilibrium equation (2.2.1):

$$\sum_{k'} p(k')d(k',k) = \sum_{k'} p(k)d(k,k').$$

The transition rates $d(k',k) = D(k',k)/T(k')$ are restricted to the following cases:

$$d(k',k) = \begin{cases} p_{ij}X_i(k_i + 1), & \text{if } k' = (k_1,\dots,k_1 + 1,\dots,k_j - 1,\dots,k_N) \\ & \text{for all } i \neq j \\ 0 \text{ in the other cases.} \end{cases} \tag{2.4.1}$$

where $k = (k_1,\dots,k_N)$, and where for all $i = 1,\dots,N$, $\sum_{j=1}^N p_{ij} = 1$.

The form which we impose on $d(k',k)$ implies that we are putting $p_{ii} = 0$, that is we do not allow customers to leave a queue and return there immediately. This restriction has already been examined in Section 2.2. In fact it is not really a restriction since it is equivalent to saying that a customer's departure from a queue which ends with his immediate return to the same queue is considered as an 'an event which did not happen'!

Let us denote a solution of the system of equations (different from system (2.2.4)) by e_i, $1 \leqslant i \leqslant N$:

$$e_i = \sum_{j=1}^N e_j p_{ji}, \quad i = 1,\dots,N. \tag{2.4.2}$$

These equations are not independent because of the constraint imposed on the sum of the p_{ij}. Their solution will not be unique.

This time, the solution of the equilibrium equations (2.2.1) for the general closed network described by the constraints (2.4.1) is:

$$p(k) = [G(K, N)]^{-1} \prod_{i=1}^{N} \prod_{m=1}^{k_i} \left[\frac{e_i}{X_i(m)} \right] \tag{2.4.3}$$

where the e_i are any solution of (2.4.2) and:

$$G(K, N) = \sum_{\substack{k \\ \sum_1^N k_i = K}} \prod_{i=1}^{N} \prod_{m=1}^{k_i} \left[\frac{e_i}{X_i(m)} \right]. \tag{2.4.4}$$

The proof is entirely analogous to those given in the preceding sections. From the equation:

$$\sum_{k'} \frac{p(k')}{p(k)} d(k', k) = \sum_{k'} d(k, k')$$

which becomes, by using (2.4.3) and (2.4.1):

$$\sum_{\substack{i,j=1 \\ i \neq j}}^{N} \frac{e_i}{X_i(k_i + 1)} \frac{X_j(k_j)}{e_j} p_{ij} X_i(k_i + 1) = \sum_{\substack{i,j=1 \\ i \neq j}}^{N} p_{ij} X_i(k_i)$$

and on simplifying one obtains:

$$\sum_{\substack{i,j=1 \\ i \neq j}}^{N} \frac{X_j(k_j)}{e_j} e_i p_{ij} = \sum_{\substack{i,j=1 \\ i \neq j}}^{N} p_{ij} X_i(k_i)$$

which can be written:

$$\sum_{j=1}^{N} \frac{X_j(k_j)}{e_j} \sum_{\substack{i=1 \\ i \neq j}}^{N} e_i p_{ij} = \sum_{i=1}^{N} X_i(k_i) \sum_{\substack{j=1 \\ j \neq i}}^{N} p_{ij}$$

or by using (2.4.2) one obtains:

$$\sum_{j=1}^{N} X_j(k_j) = \sum_{i=1}^{N} X_i(k_i)$$

which completes the proof of (2.4.3).

2.5 - ALGORITHMS FOR CALCULATING THE NORMALIZATION CONSTANT G(K, N) FOR CLOSED NETWORKS

The closed models which we have presented in Sections 2.3 and 2.4 require calculation of the normalization constant $G(K, N)$ to enable them to be used numerically.

One is tempted to comment that for the system of equations (2.2.1) which allows the $p(k)$ to be calculated, a direct numerical method would be sufficient since (2.2.1) is a finite system of linear equations for

$$\sum_{1}^{N} k_i = K \quad \text{constant.}$$

But this is practically impossible; as evidence it is sufficient to comment that even for small values of K and N, the number of states k and hence equations is prohibitive. In fact it can be shown that the possible number of states is:

$$\binom{N+K-1}{N-1}.$$

To show this, note that any $k = (k_1, \ldots, k_N)$ can be represented by a string of binary elements:

$$\underset{\leftarrow k_1 \rightarrow}{1\ldots1}\ \ 0\ \ \underset{\leftarrow k_2 \rightarrow}{1\ldots1}\ \ 0\ldots0\ \ \underset{\leftarrow k_N \rightarrow}{1\ldots1}$$

consisting of k_1 ones, followed by a zero, then by k_2 ones, then a zero, etc., until a last zero is followed by k_N ones. There is therefore an isomorphism between the array of strings of characters and the array of states k. But a string such as this is determined by the placing of the $(N-1)$ zeros among the $(N+K-1)$ characters (there are K ones in total and $N-1$ zeros), from which the result follows.

Even for reduced values of N and K, this number is extremely large. For example, for a small system with $N = 10$ and $K = 6$, we have 5005 states! This makes direct calculation of $p(k)$ from the equilibrium equations (2.2.1) practically impossible. It is not the same as calculation of $G(K, N)$ by summation over all states from (2.4.4).

2.5.1. Case where the X_i do not depend on the k_i

To simplify the presentation of the algorithms for calculating the normalization constant $G(K, N)$, we start with the case where:

$$X_i(m) = X_i, \quad \text{for all} \quad 1 \leqslant i \leqslant N \quad \text{and} \quad m \geqslant 0.$$

Let us put:

$$\rho_i = e_i / X_i,$$

which allows us to write:

$$G(K, N) = \sum_{\substack{k_j \geqslant 0 \\ \sum_1^N k_i = K}} \prod_{i=1}^{N} (\rho_i)^{k_i}. \tag{2.5.1}$$

We shall define the following generating functions:

$$g_i(z) = \sum_{k_i = 0}^{\infty} (\rho_i z)^{k_i} = 1/(1 - \rho_i z), \quad 1 \leqslant i \leqslant N, \tag{2.5.2}$$

and

$$g(z) = \prod_{i=1}^{N} g_i(z), \tag{2.5.3}$$

assuming that z is chosen in such a way that each series (2.5.2) converges.

Comment $G(K, N)$ is the coefficient of z^K in the series $g(z)$. We shall define the partial products:

$$\begin{cases} h_1(z) = g_1(z), \\ h_\ell(z) = h_{\ell-1}(z)g\ell(z), \quad N \geqslant \ell > 1. \end{cases} \tag{2.5.4}$$

By using (2.5.2), we obtain:

$$h_\ell(z) = h_{\ell-1}(z) + \rho_\ell z h_\ell(z) \tag{2.5.5}$$

In (2.5.4) it is evident (see the comment above) that $G(j, \ell)$ is the coefficient of z^j in the series $h_\ell(z)$ where $G(j, \ell)$ is obtained by replacing K by j and N by ℓ in (2.5.1). We can therefore exploit this fact in (2.5.5) to write that, for

$$N \geqslant \ell > 1, \, K \geqslant j \geqslant 1: \tag{2.5.6}$$
$$G(j, \ell) = G(j, \ell - 1) + \rho\ell\, G(j - 1, \ell).$$

But since:

$$G(j, 1) = \rho_1^j, \quad K \geqslant j \geqslant 1,$$

the equality (2.5.6) can be used to calculate $G(K, N)$ effectively, and therefore provides a simple algorithm for the purpose. In algorithmic language it can be written:

Procedure 1   *integer* N; *integer* K;
real [RHO(L): $1 \leqslant L \leqslant N$];
real [G(J, L): $1 \leqslant J \leqslant K$, $1 \leqslant L \leqslant N$];

for J from 1 to K

$$G(J, 1) \leftarrow RHO(L)**J$$

end for
  for J from 1 to K:
    for L from 2 to N:

$$G(J, L) \leftarrow 0$$

    end for
  end for
  for J from 1 to K:
    for L from 2 to N:

$$G(J, L) \leftarrow G(J, L - 1) + RHO(L)*G(J - 1, L)$$

    end for
  end for
  end for
end procedure 1

This algorithm requires about N.K arithmetic operations. This compares with the $(N + K - 1)!/(N - 1)!K!$ additions necessary to calculate $G(K, N)$ directly from (2.5.1), without counting the multiplications.

2.5.2. Algorithm for calculating $G(K, N)$ in the general case

When the simplifying assumption $X_i(m) \equiv X_i$ is not observed, the algorithm for calculating the normalization constant $G(K, N)$ from formula (2.4.3) becomes slightly more complicated. Let us write the generating function associated with the network again:

$$g(z) = \prod_{i=1}^{N} g_i(z)$$

where this time:

$$g_i(z) = \sum_{k_i=0}^{\infty} z^{k_i} \prod_{m=1}^{k_i} [e_i/X_i(m)]. \tag{2.5.7}$$

As in Section 2.5.1, we notice that $G(K, N)$ is the coefficient of z^K in $g(z)$. We again use the partial products:

$$h_1(z) = g_1(z),$$
$$h_i(z) = h_{i-1}(z)g_i(z), \quad 2 \leqslant i \leqslant N,$$

to calculate the $G(j, i)$. The preceding recurrence indicates, in particular, that $G(j, i)$ is the coefficient of z^j in $h_i(z)$ and that it can be written:

$$G(j, i) = \sum_{\ell=0}^{j} G(\ell, i - 1) \prod_{m=1}^{j-\ell} [e_i/X_i(m)],$$

for $2 \leqslant i \leqslant N$, with:

$$G(j, 1) = \prod_{m=1}^{j} [e_1/X_1(m)] \quad \text{for all } j,$$

and
$$G(0, i) = 1 \quad \text{for } 1 \leqslant i \leqslant N.$$

Let us put:
$$\rho_i(j) = \prod_{m=1}^{j} [e_i/X_i(m)].$$

In algorithmic language, this procedure can be written in the following manner:

Procedure 2
        real $[E(I):1 \leqslant I \leqslant N]$
        real $[X(I, M):1 \leqslant I \leqslant N, 1 \leqslant M \leqslant K]$
        real $[G(J, I):0 \leqslant J \leqslant K, 1 \leqslant I \leqslant N]$
        real $[RHO(J, I):0 \leqslant J \leqslant K, 1 \leqslant I \leqslant N]$
for I from 1 to N: $G(0, I) \leftarrow 1$; $RHO(0, I) \leftarrow 1$;
    for J from 1 to K:
    for M from 1 to J:

$$\text{RHO}(M, I) \leftarrow \text{RHO}(M - 1, I)*X(I, M)$$

    end for
   RHO(J, I) ← RHO(J, I)*(E(I)**J)
   end for
  end for
for J **from** 1 **to** K:G(J, 1) ← RHO(J, 1) **end for**
for I **from** 1 **to** N − 1:
  for J **from** 1 **to** K:
   for L **from** 0 **to** J:
   G(J, I + 1) ← G(J, I + 1) + G(L, I)*RHO(J − L, I + 1)
   end for
  end for
 end for
end procedure 2

This procedure requires a number of arithmetic operations proportional to NK^2.

2.5.3. Calculation of the utilization rates of each server

The rate of utilization of server i, which we denote by U_i, is defined by:

$$U_i(K, N) = \sum_{\substack{k_i > 0, k_j \geqslant 0 \\ j = 1, \ldots, N \\ j \neq i}} p(k).$$

That is the sum of the $p(k)$ corresponding to all states k such that $k_i > 0$, the other k_j can take the values $0, \ldots, K$ with the usual constraint for a closed network: $\sum_1^N k_j = K$. From $g(z)$, one can write:

$$\tilde{g}_i(z) = g(z) \frac{g_i(z) - 1}{g_i(z)}.$$

That is to say that $\tilde{g}_i(z)$ is obtained by replacing $g_i(z)$ by $g_i(z) - 1$, which means the smallest value taken by k_i in $\tilde{g}_i(z)$ is $k_i = 1$. Hence $G(K, N)U_i(K, N)$ is the coefficient of z^k in $\tilde{g}_i(z)$.

In a slightly different form one has:

$$\tilde{g}_i(z) = g(z)[1 - (g_i(z))^{-1}]$$

and hence:

$$G(K, N)U_i(K, N) = G(K, N) - \sum_{\ell=0}^{k} G(\ell, N)\theta_i(K - \ell)$$

where $\theta_i(\ell)$ is the coefficient of z in $(g_i(z))^{-1}$; the term $\theta_i(\ell)$ can then be determined from the relation:

$$g_i(z)(g_i(z))^{-1} = 1.$$

Hence one obtains for $1 \leqslant i \leqslant N$:

$$\theta_i(0) = 1$$

and

$$\sum_{\ell=0}^{j} \theta_i(\ell)\rho_i(j-\ell) = 0.$$

Since $\rho_i(0) = 1$, we can write:

$$\theta_i(j) = -\sum_{l=0}^{j-1} \theta_i(\ell)\rho_i(j-\ell)$$

which suggests an algorithm for calculating the $\theta_i(j)$ and hence equally the $U(K, N)$.

In the particular case treated in Section 2.5.1 where:

$$\rho_i(j) = (e_i/X_i)^j = \rho_i^j,$$

we shall have:

$$(g_i(z))^{-1} = 1 - \rho_i z,$$

and thus:

$$\theta_i(j) = \begin{cases} 1 & \text{if } j = 0, \\ -\rho_i & \text{if } j = 1, \\ 0 & \text{for } j > 1. \end{cases}$$

This results in:

$$G(K, N)U_i(K, N) = G(K, N) - G(K, N) + \rho_i G(K - 1, N),$$

that is for $1 \leqslant i \leqslant N$:

$$U_i(K, N) = \rho_i G(K - 1, N)/G(K, N).$$

In this last case, the algorithm for calculating $G(K, N)$ given in Section 2.5.1 allows that $U_i(K, N)$ to be easily obtained also.

2.6 - JACKSON'S THEOREM

In this section, we present the results obtained by Jackson for queueing networks with exponential service times and Poisson arrival distributions. We treat open systems (that is with users arriving from and departing to the exterior of the network) separately from closed systems. In a system described as closed, the number of users circulating from one service station to another is held constant.

2.6.1. Open network: Chapman–Kolmogorov equation

Let us start by studying the open network represented in Figure 2.2 in which the queues are interconnected in any manner. Customers arrive from the exterior of the network following a Poisson process of parameter λ; hence the time intervals between successive arrivals are random variables distributed according to an exponential distribution. The network contains N servers, each one

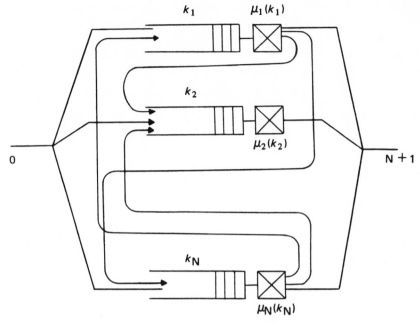

Figure 2.2 Open Jackson network

having a queue. Customers arriving from the exterior move towards the ith queue with a constant probability p_{0i}. That is equivalent to a system in which each queue is 'fed' from the exterior of the network following a Poisson debit process λp_{0i} with:

$$\sum_{i=1}^{N} p_{0i} = 1, \quad p_{0i} \geqslant 0.$$

Let $p_{i,j}$, $1 \leqslant i \leqslant n$, $1 \leqslant j \leqslant N + 1$, be the constant probability that a customer, having finished being served by server i, goes to the queue of server j, $1 \leqslant j \leqslant N$, or to the exterior if $j = N + 1$. The service times of customers with the same server or with different servers will be independent random variables. The service time of server i when k_i customers (including the one receiving service) are in his queue is a random variable distributed according to an exponential law of parameter $0 < \mu_i(k_i) < \infty$. Such a server will henceforth be called a *Jacksonian server*.

Let, as before, $k = (k_1, \ldots, k_N)$, and let $k(t)$ be the random vector formed from the number of customers $k_i(t)$ at each of the servers $1 \leqslant i \leqslant N$ at time t. The differential and difference equations describing the evolution of the probability distribution $p(k, t)$ are obtained in the same manner as for the system which we studied in Section 3.3, where

$$p(k, t) = P\{\bar{k}(t) = k \,|\, \bar{k}(0) = k^0\}, \tag{2.6.1}$$

k^0 being a given initial condition. The following notation is identical to that used in Section 2.2.

Let $a(k, i)$ and $b(k, i)$ be vectors which are identical to the vector k except that component k_i is replaced by $k_i + 1$ and $k_i - 1$ respectively. Vector $c(k, i, j)$ is identical to k with the exception of components k_i and k_j which are replaced by $k_i + 1$ and $k_j - 1$.

The system equations, called the 'Chapman–Kolmogorov equations', are obtained in the same way as in Section 1.3.3 and can be written for each k:

$$\frac{d}{dt} p(k, t) = -\left(\lambda + \sum_{i=1}^{N} \mu_i(k_i)(1 - p_{ii}) \right) p(k, t)$$

$$+ \sum_{i=1}^{N} \lambda p_{0i} p(b(k, i), t)$$

$$+ \sum_{i=1}^{N} \mu_i(k_i + 1) p_{i, N+1} p(a(k, i), t) \qquad (2.6.2)$$

$$+ \sum_{\substack{i,j=1 \\ i \neq j}}^{N} \mu_i(k_i + 1) p_{ij} p(c(k, i, j), t)$$

where we simplify the notation by assuming that $\mu_i(k_i) = 0$ if k_i is zero and that $p(k, t) = 0$ if the vector k contains a negative component.

Equations (2.6.2) can be considered as an infinite system of linear differential equations of which the matrix has constant coefficients, $E = [E_{kk'}]$.

$$\frac{d}{dt} p(k, t) = \sum_{k'} E_{kk'} p(k', t).$$

We define the stationary solution for system (2.6.2), if it exists, as the probability distribution $\{p(k)\}$ independent of the initial conditions and given by:

$$p(k) = \lim_{t \to \infty} p(k, t). \qquad (2.6.3)$$

In general, this limit may not exist. If it exists, it will be obtained by putting:

$$\frac{d}{dt} p(k, t) = 0$$

in (2.6.2). For a rigorous study of the conditions for the existence of this solution, we refer the reader to mathematical works such as that of Feller [8].

2.6.2. Open network: use of the method of points of regeneration

Instead of studying the stationary solution $\{p(k)\}$ of the system of equations (2.6.2) by using condition (2.6.3), we are going to make use (as in Chapter 1) of the regenerative structure of the process $\{\bar{k}(t), t \geq 0\}$ due to the fact that the arrivals are Poisson.

Let us denote the vector for which all the elements are zero by $\underline{0} =$

$(0,\dots,0)$ and consider the series of instants $t_0 < t_1 < \cdots < t_n < \cdots$ for which:

$$\overline{k}(t_n) \neq \underline{0}, \quad \overline{k}(t_n) = \underline{0}, \quad n \geq 0,$$

if such instants exist. Taking account of the assumptions made, one could prove that for all $\overline{k}(0) = k^{-0}$ (initial condition), $E(t_0) < \infty$, since the process $\{\overline{k}(t), t \geq 0\}$ is an irreducible and aperiodic Markov chain. However, we shall simplify the problem by assuming that $t_0 = 0$.

Since the network becomes empty at time t_n, and the arrival process is Poisson, then for each n, $\{\overline{k}(t), t \geq t_n\}$ is independent of $\{\overline{k}(t), t < t_n\}$; on the other hand, the behaviour of $\{\overline{k}(t - t_n), t \geq t_n\}$ is identical, for all probability distributions which govern it, to that of $\{\overline{k}(t), t \geq 0\}$. Hence the $\{t_n, n \geq 1\}$ are the regeneration points of the process.

Let us put

$$T_n(k) = \int_{t_{n-1}}^{t_n} 1(\overline{k}(t) = k)\,dt, \tag{2.6.4}$$

for $n \geq 1$, and hence:

$$T_n = (t_n - t_{n-1}) \tag{2.6.5}$$

which is

$$T_n = \sum_n T_n(k).$$

We know that the stationary probability $p(k)$ can be obtained by (see Chapter 1, Section 3.2):

$$p(k) = \frac{ET_n(k)}{ET_n}, \quad \text{for any } n. \tag{2.6.6}$$

In the following we assume that the value of n is fixed and we use (2.6.6) without specifying it each time:

$$p(k) = \frac{ET(k)}{ET}.$$

To calculate $p(k)$ we proceed as before in Section 2.2.

Theorem 2.6.1

For the open Jackson network, the stationary probability distribution, if it exists, is given by:

$$p(k) = p(\underline{0}) \prod_{i=1}^{N} \prod_{m=1}^{k_i} (\lambda e_i / \mu_i(m)) \tag{2.6.7}$$

where the e_i are the solution of the system equations:

$$e_i = p_{0i} + \sum_{j=1}^{N} e_j p_{ji}, \quad i = 1,\dots,N,$$

and where:

$$p(\underline{0}) = \left(\sum_k \prod_{i=1}^{N} \prod_{m=1}^{k_i} \frac{\lambda e_i}{\mu_i(m)} \right)^{-1}.$$

The necessary and sufficient condition for existence is $p(0) > 0$.

Comment The form of the solution (2.6.7) of the Jackson network is analogous to that (2.2.2) obtained previously for the deterministic open network. The analogy between these two systems will be used to prove (2.6.7). Also, e_i is the *mean number of visits* made by any customer to station i ($1 \leqslant i \leqslant N$) while in the queueing network.

Proof For a given state k, we can write the following relation:

$$ET(k) = EM(k) \cdot E\tau(k)$$

where $EM(k)$ is the mean number of times that state k is observed during the interval, and $E\tau(k)$ is the mean time spent in state k during a visit.

In fact, $E\tau(k)$ is easily obtained since the service times and interarrivals obey exponential distributions

$$E\tau(k) = \left[\lambda + \sum_{i=1}^{N} \mu_i(k_i) \right]^{-1}.$$

The $EM(k)$ must satisfy the relation:

$$EM(k) = \sum_{(i,j)} \pi_{c(k,i,j),k} M[c(k,i,j)] + \sum_i \pi_{b(k,i)k} M[b(k;i)]$$

$$+ \sum_i \pi_{a(k,i),k} M[a(k,i)]. \tag{2.6.8}$$

where $\pi_{kk'}$ is the probability of passing from state k to state k'. Due to the memoryless property of the exponential laws of interarrivals and service durations, the preceding transition probabilities are easily calculated and from (2.6.8) one obtains:

$$EM(k) = \sum_{(i,j)} \frac{\mu_i(k_i + 1)p_{ij}}{\lambda + \sum_j \mu_j(k_j)} M[c(k,i,j)]$$

$$+ \sum_i \frac{\lambda p_{0i}}{\lambda + \sum_j \mu_j(k_j)} M[b(k,i)] \tag{2.6.9}$$

$$+ \sum_i \frac{\mu_i(k_i + 1)p_{i,N+1}}{\lambda + \sum_j \mu_j(k_j)} M[a(k,i)]$$

By using the relations $ET(k) = EM(k) \cdot E\tau(k)$ and dividing the two terms by ET (assuming ET to be finite), one obtains the relation (2.2.1):

$$\sum_{k'} p(k') d(k',k) = \sum_{k'} p(k) d(k,k')$$

for which the solution (see Section 2.2) is obtained by replacing X_0 by λ and $X_i(m)$ by $\mu_i(k_i)$, from which the stationary probability distribution (2.6.7) follows.

2.6.3. Closed networks

In this section we are interested in the steady state behaviour of a *closed* queueing network. Consider a system of N servers, each with a queue, interconnected in any way and assume that a constant number of customers K circulate between the different service stations. In this network there are neither arrivals from nor departures to the exterior.

We shall use the same notation as for the open network:

$$k = (k_1, k_2, \ldots, k_N)$$

is the vector giving the number of customers k_i at each of the N servers ($1 \leqslant i \leqslant N$).

Taking account of the assumptions made, the system is completely described by the probability distribution of the vectors k at the instant t, $\{p(k, t)\}$, which must satisfy the following equations:

$$\frac{d}{dt} p(k, t) = - \sum_{i=1}^{N} \mu_i(k_i)(1 - p_{ii})p(k, t)$$

$$+ \sum_{\substack{i,j=1 \\ i \neq j}}^{N} \mu_i(k_i + 1)p_{ij}p(c(k, i, j), t) = 0 \qquad (2.6.10)$$

which differ from (2.6.2) only by the absence of terms representing the exchange of customers with the exterior.

Notice that the components k_i of the vector k are such that:

$$\sum_{i=1}^{N} k_i = N \quad \text{and} \quad k_i \geqslant 0, 1 \leqslant i \leqslant N,$$

and hence equation (2.6.10) constitutes a *finite* system of linear differential equations with constant coefficients which can be written in the form:

$$\frac{d}{dt} Q(t) = HQ(t) \qquad (2.6.11)$$

where $Q(t)$ is the vector $\{p(k, t)\}$ and H the coefficient matrix. It can be shown that this matrix has a zero eigen value and that all other eigen values have negative real parts. It follows that the stationary distribution defined by $p(k) = \lim_{t \to \infty} p(k, t)$ exists.

Theorem 2.6.2

For a closed Jackson network, the probability distribution is given by:

$$p(k) = p(\underline{0}) \prod_{i=1}^{N} \prod_{m=1}^{k_i} [e_i/\mu_i(m)] \qquad (2.6.12)$$

where the e_i are any *one* solution to the system of equations

$$e_i = \prod_{j=1}^{N} e_j p_{ji}, \quad i = 1, \ldots, N,$$

and where

$$p(\underline{0}) = \left[\sum_{k_1 + \cdots + k_N} \prod_{i=1}^{N} \prod_{m=1}^{k_i} [e_i/\mu_i(m)] \right]^{-1}.$$

Comment The system of equations $e_i = \sum_{j=1}^{N} e_j p_{ji}$ does not have a unique solution since the determinant of the matrix (p_{ij}) is zero. However, any solution to this system of equations, in particular that obtained by putting $e_1 = 1$, can be substituted in (2.6.12). In fact, $p(k)$ does not depend on the chosen solution e_i, $1 \leqslant i \leqslant N$. If one takes $e_1 = 1$, $e_i(i > 1)$ becomes the mean number of visits by any client to station i relative to the number of visits to station 1.

This theorem can be proved directly from the Chapman–Kolmogorov equations or from the regenerative structure of the process $\{k(t), t \geqslant 0\}$. The proof is virtually identical to that of Section 2.6.2.

Notice an important point: formally the calculation of $p(0)$ for a closed network with N stations and K customers is identical to that of $[G(K, N)]^{-1}$ given by equation (2.4.4). The algorithms of Section 2.5 can therefore be used if $X_i(m)$ is replaced by $\mu_i(m)$.

2.6.4. Application example

By way of example, we shall apply theorem 2.6.2 to the closed network of Figure 2.3. A system with three service stations is represented there; the service

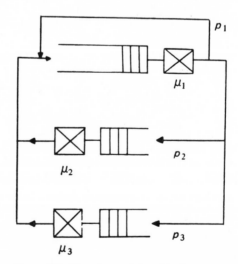

Figure 2.3 Closed network with three service stations

times are independent random variables with distributions following exponential laws with means of $1/\mu_1$, $1/\mu_2$ and $1/\mu_3$ respectively.

Let K be the total number of customers in the system. Let us call p_2 and p_3 the probability that a customer who has been served by server 1 goes to server 2 or 3 respectively, and

$$p_1 = 1 - (p_2 + p_3)$$

is the probability that this customer returns directly to the queue at station 1. The matrix $P[p_{ij}]$ becomes:

$$P = \begin{bmatrix} p_1 & p_2 & p_3 \\ 1 & 0 & 0 \\ 1 & 0 & 0 \end{bmatrix}.$$

The solution of the system $e = eP$, taking $e_1 = 1$, is:

$$e_1 = 1,$$
$$e_2 = p_2,$$
$$e_3 = p_3.$$

As the system is closed, the assumptions of the theorem are always satisfied and we obtain:

$$p(k_1, k_2, k_3) = (G'(K))^{-1} \left(\frac{1}{\mu_1} \right)^{k_1} \left(\frac{p_2}{\mu_2} \right)^{k_2} \left(\frac{p_3}{\mu_3} \right)^{k_3}.$$

Multiplying the expression on the right by μ_1^K, we have:

$$p(k_1, k_2, k_3) = \frac{1}{G'(K)} \left(\frac{p_2 \mu_1}{\mu_2} \right)^{k_2} \left(\frac{p_3 \mu_1}{\mu_3} \right)^{k_3}.$$

The normalization constant $G'(K)$ is obtained very easily:

$$G'(K) = \sum_{k_1 + k_2 + k_3 = K} \left(\frac{p_2 \mu_1}{\mu_2} \right)^{k_2} \left(\frac{p_3 \mu_1}{\mu_3} \right)^{k_3}.$$

By applying the results of Section 2.5.3, the occupation rate of server 1 can be written in the following manner:

$$U_1 = \sum_{\substack{k_1 > 0 \\ k_1 + k_2 + k_3 = K}} p(k_1, k_2, k_3) = \frac{G'(K - 1)}{G'(K)}.$$

2.7 - APPLICATION TO ANALYSIS OF THE CHARACTERISTICS OF A COMPUTER WITH VIRTUAL MEMORY

The model of a partitioned system with virtual memory is represented in Figure 2.4. It contains three servers: the central processing unit, the secondary memory (or page drum) and the file disc which models the inputs–outputs.

Programs entering the system are put in a queue for the central processing

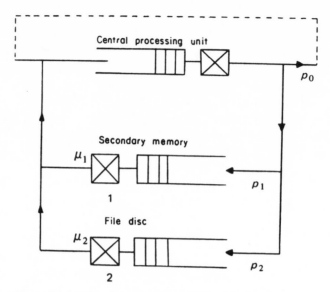

Figure 2.4 Model of a computer system with virtual memory

unit. Execution of a program on the central processing unit continues until an interrogation caused by one of the following events:

–end of execution; the total calculation time of a program finishes. At the end of this time, the program leaves the system and is replaced by a new program. This is represented by the branch p_0 in Figure 2.4;
–a page shortage, in which case the program is put in a queue for the secondary memory;
–an input–output; in this case the program is put in a queue for the file disc.

Service demands for the secondary memory and the file disc are treated in first arrived, first served order. Operation of the units is represented by the total time demanded for a service. We assume that these times are exponentially distributed with respective means of μ_1^{-1} and μ_2^{-1}.

The behaviour of a program, determined by a set of parameters, controls the running of a service on the central processing unit. In our model, all possible behaviours are characterized by the distributions of three variables which are:

–the total calculation time of a program;
–the total calculation time between two input–output demands in the central processing unit;
–the total calculation time on the central processing unit between two page shortages.

The first two distributions depend only on the program considered. The third depends equally on the system state and, in particular, the central memory space assigned to the program. This last law, assumed to be exponential, has a mean q given by Belady's formula:

$$q = as^u$$

where s is the portion of the memory allocated to programs, u is a coefficient depending on the program which we shall call its locality and a is a coefficient depending on the machine, the system (the nature of the control algorithm in the central memory) and the program. Let M be the total memory reserved for program execution and K the degree of multiprogramming which is, by definition, the number of programs in the system at a given instant. We assume that the memory is equally shared between all programs, that is, we have:

$$s = \frac{M}{K}, \quad \text{from which} \quad q = a\left(\frac{M}{K}\right)^u.$$

We take $a = 0.01$ and $u = 1.5$ or $u = 2.5$. The distributions followed by the total calculation time and the times between two input–output demands are assumed to be exponential with means of c and r. The parameters of a program are:

−its total calculation time, determined randomly,
−its mean calculation time between input–outputs r,
−its locality u.

Before resolving the system, let us show that the central processing unit is an exponential server, that is an uninterrupted service time t on the central processing unit is the minimum of the three times t_1, t_2 and t_3 of the exponential distributions having the following distribution functions respectively:

$F_1(t_1) = 1 - e^{-t_1/c}$ for the time t_1 before the end of the calculation,
$F_2(t_2) = 1 - e^{-t_2/q}$ for the time t_2 before a page fault,
$F_3(t_3) = 1 - e^{-t_3/r}$ for the time t_3 before an input–output demand.

Let $F(t)$ be the distribution function of the uninterrupted service times t. One has:

$F(t) = P\{\tau \leqslant t\}$ with $\tau = \min(t_1, t_2, t_3)$ from which
$F(t) = 1 - P\{t_0 > t\}P\{t_1 > t\}P\{t_2 > t\}$ from which
$F(t) = 1 - [1 - F_1(t)][1 - F_2(t)][1 - F_3(t)]$ hence

$$F(t) = 1 - \exp\left[-t\left(\frac{1}{c} + \frac{1}{q} + \frac{1}{r}\right)\right] = 1 - e^{-\mu_0 t},$$

by putting

$$\mu_0 = \frac{1}{c} + \frac{1}{q} + \frac{1}{r}.$$

Let p_0, p_1 and p_2 be the probabilities of exit from the system, page fault and input–output demand respectively for the program at the end of service by the

68

central unit. We have:

$$p_1 = \int_0^\infty P\{(t - dt \le t_1 \le t) \quad \text{and} \quad (t_0 > t) \quad \text{and} \quad (t_2 > t)\}$$

for which

$$p_1 = \int_0^\infty \frac{1}{q} e^{-\mu_0 t} \, dt = \frac{1}{\mu_0 q}.$$

Similarly:

$$p_2 = \frac{1}{\mu_0 r},$$

from which

$$p_0 = 1 - p_1 - p_2 = \frac{1}{\mu_0 c}.$$

Utilization rate of
the central processing unit A_0 (K)

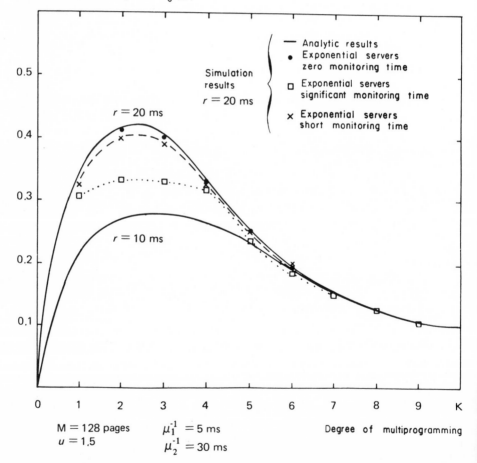

Figure 2.5 Utilization rate of the central processing unit ($u = 1.5$)

The model clearly verifies Jackson's assumptions of Section 2.6 and resolves itself as a stationary system as in the example of Section 2.6. Let $p(k_0, k_1, k_2)$ be the probability that there are k_0 customers in the queue for the central processing unit, k_1 in that for the secondary memory and k_2 in that for the file disc and let A_0, A_1 and A_2 be the utilization rates of these three units. As in the solution of the preceding example, we have:

$$p(k_0, k_1, k_2) = \frac{1}{G(K)} \left(p_1 \frac{\mu_0}{\mu_1} \right)^{k_1} \left(p_2 \frac{\mu_0}{\mu_2} \right)^{k_2}.$$

Utilization rate of
the central processing unit $A_0(K)$

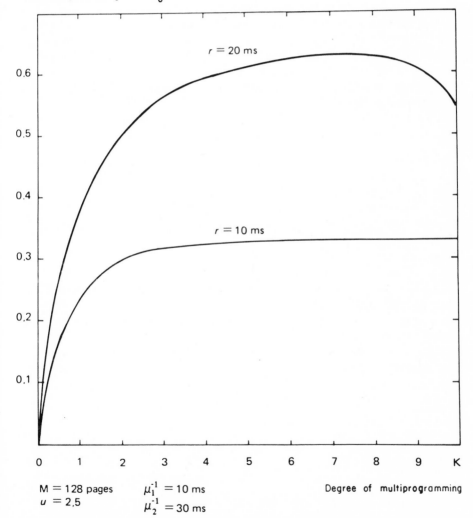

M = 128 pages μ_1^{-1} = 10 ms   Degree of multiprogramming
u = 2,5 μ_2^{-1} = 30 ms

Figure 2.6 Utilization rate of the central processing unit ($u = 2.5$)

Since $A_0(K) = \sum_{1 \leqslant k_0 \leqslant K} P(k_0, k_1, k_2)$ after summation we have:

$$A_0(K) = \frac{G(K-1)}{G(K)}.$$

We also have:

$$A_1(K) = \frac{1}{\mu_1 q} A_0(K), \quad A_2(K) = \frac{1}{\mu_2 r} A_0(K).$$

The results obtained are presented in Figures 2.5 and 2.6 in which $M = 128$, $u = 1.5$ or 2.5, $r = 10$ milliseconds (ms) or 20 ms, $1/\mu_1 = 5$ ms, $1/\mu_2 = 30$ ms and $c = 800$ ms. They give the value of the utilization rate of the central processing unit as a function of the degree of multiprogramming K.

The principal conclusion to be drawn from the curves of Figures 2.5 and 2.6 is that the system collapses when the degree of multiprogramming is too large: more and more users in the system are allocated less and less memory and cause more and more page faults. The onset of the phenomenon of collapse as K increases is determined essentially by the locality u of the programs and by the service times $1/\mu_1$ of the secondary memory. A high locality u is favourable and a good feature of the system.

EXERCISES

1. Find the solution for the M/M/1 system with feedback, as represented in the following.

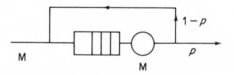

This system can represent an output line of a switching node whose task is to transmit messages to another node. If the message is transmitted with probability p, it is correctly transmitted. If it is transmitted with probability $1 - p$, it is erroneous and must be retransmitted.

$(1 - p)$ can represent the probability of a transmission error necessitating retransmission of the message; hence the messages constitute the customers of the queue.

Find in particular the maximum output from this output line.

2. Study the system shown in the following figure.

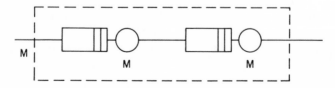

The total number of customers in the system is limited to K. The service times are exponentially distributed with means of μ_1^{-1} and μ_2^{-1}. Arrivals from the exterior form a Poisson process.

3. Consider the system described in the following figure. Find the maximum value of λ which gives rise to a stable system.

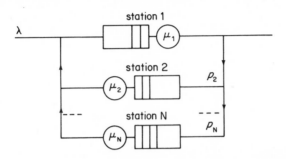

Assuming that the total number of customers in the system is limited to $k_1 + \cdots + k_N \leqslant K$ where k_j is the number of customers in queue i, find the maximum output from the system (it is necessary to assume that each time a customer leaves a new one enters, this returns to the study of a closed network with K customers).

4. Consider a network $\mathcal{R}$ to which Jackson's theorem is applicable. It can be shown that streams which go to the exterior of the network are Poisson processes. Let the equivalence be defined by $i \sim j$ if, and only if, one can gain access to queue j from queue i and to queue i from queue j.

Determine the equivalence classes which we shall call irreducible subnetworks: $\mathcal{R}_i, i = 1, \ldots, n$ such that $\cup \mathcal{R}_i = \mathcal{R}$. Where $\mathcal{R}_0$ and $\mathcal{R}_{N+1}$ are two fictitious stations representing the input and output of the network.

From the result stated at the beginning, show that the streams between two irreducible sub-networks form a Poisson process. Furthermore, it is possible to show that all the streams from the interior of an irreducible sub-network are neither Poisson nor renewal processes.

Conclusion: In a Jackson network, the only Poisson streams are those which link $\mathcal{R}_i$ and $\mathcal{R}_j, i, j \in 0, 1, \ldots, N+1, i \neq j$.

5. Study the M/G/1/K system and show that:

$$p_k(n) = \begin{cases} c_k \hat{p}(n), & 0 \leqslant n < K, \\ 1 - [1 - c_k(1 - \rho)]/\rho, & n = K, \end{cases}$$

where

$$\rho = \frac{\lambda}{\mu} \quad \text{and} \quad c_k = \left\{ 1 - \rho \left[1 - \sum_{n=0}^{K-1} \hat{p}(n) \right] \right\}^{-1}$$

and $\hat{p}(n)$ is the probability of having n customers in the M/G/1/∞ system.

Answer

It is first necessary to show that the M/G/1/K system is equivalent to the closed system represented below.

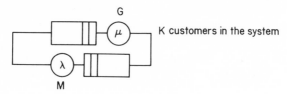

From which $p(n, K - n) = P\{n$ customers in queue 1 and $K - n$ in queue 2$\}$ $= p_k(n)$.

BIBLIOGRAPHY

The results presented in this chapter are inspired by Jackson's paper [1], where the preceding partial results concerning exponential queues are cited. The use of these models, and in particular the model with a central server (Sections 2.3 and 2.6.4), for the analysis of computer systems has been introduced by Buzen [2]; a more complete presentation of these applications can be found in [3]. The deterministic approach presented in Sections 2.2 and 2.4 was developed initially in [4], a synthesis appears in [5], and the algorithms for calculating the normalization constants (Section 2.5) appear in [2] and [6]. The model of Section 2.7 was presented in [7].

The proofs for Jackson's results proposed here are different from those normally used.

1. Jackson, J. R. (1963). Jobshop like queueing systems, *Management Science*, **10**, 131–142.
2. Buzen, J. P. (1973). Computational algorithms for closed queueing networks with exponential servers, *Comm. ACM*, **16**, 527–531.
3. Gelenbe, E. and Mitrani, I. (1980). Analysis and Synthesis of computer System Models, Academic Press, London and New York.
4. Buzen, J. P. (1976). Fundamental operational laws of computer performance, *Acta Informatica*, **7**, 167–182.
5. Denning, P. J. and Buzen, J. P. (1978). The operational analysis of queueing network model models, *ACM Computing Surveys*, **10**, 225–262.
6. Williams, A. C. and Bhandiwad (1976). A generating function approach to queueing network analysis of multiprogrammed computers, *Networks*, **6**, 1–22.
7. Badel, M., Gelenbe, E., Leroudier, J. and Potier, D. (1975). Adaptive control of virtual memory systems, *Proceedings IEEE*.
8. Feller, W. An Introduction to Probability Theory and its Application, Vol. 1, John Wiley, Princeton, NJ.

CHAPTER 3

Extensions to Queues with
A Single Server

In Chapter 1 we studied queues containing only a single server; the capacity of the queue was unlimited and the service discipline was first come, first served. Also, the distribution of service times did not depend on the number of customers present in the queue. In this chapter we are going to abandon these assumptions in order to study more general queues. Firstly, we are going to study queues with Poisson arrivals and exponential service in which the number of servers can be greater than 1, the global capacity of the queue can be limited and the service discipline can be different from first come, first served. Secondly, we shall abandon the assumptions which make the queue Markovian. To obtain a solution we shall make particular use of a diffusion approximation. Finally, the case involving several classes of customers will be tackled at the end of the chapter.

3.1 - EXTENSIONS TO THE M/M/1 QUEUE

3.1.1 M/M/1 queue with state dependence

The queue considered in this chapter is characterized by an arrival process and a service time distribution which depend on the number of customers in the queue.

When there are n customers in the queue (counting equally the customer being served) the arrival process is Poisson with parameter $\lambda(n)$ and the service time is exponentially distributed with rate $\mu(n)$. It should now be mentioned that we can define such a queue only because the distributions are exponential. In fact, since this law is memoryless, it is possible to start the service again (or to continue, which is equivalent) when a customer with a new parameter $\mu(n + 1)$ enters the system. In contrast, if the intervals between two arrivals or the service times follow general distributions it is essential to take account of the amount of service already provided. It will be necessary, for example, in order to analyse this last queue, to fix the service time at the moment when the customer arrives at the service point and not to change it again. In the M/M/1 case which interests us, the service time is determined in reality by the number of customers in the queue just before a departure or an arrival.

Definition A birth and death process is a Markov process in which only two transitions are possible: the process decreases by 1 (a death) or the process increases by 1 (a birth).

Let N_t be the number of customers at time t.

Theorem

N_t is a birth and death process.

Proof We shall use the property of exponential laws and the independence of arrivals and services. Let V_t and W_t be the time intervals starting at t and finishing with a birth or a death respectively. We have:

$$P\{V_t > t, W_t > t | N_s = i \quad \text{and} \quad s \leqslant t\} = e^{-(\lambda(i) + \mu(i))t}$$

and

$$P\{V_t \leqslant W_t | N_s = i \quad \text{and} \quad s \leqslant t\} = \frac{\lambda(i)}{\lambda(i) + \mu(i)},$$

which clearly shows that N_t is a Markov process.

$\lambda(i)$ and $\mu(i)$ are the birth and death rates when the population of the queue is i customers.

The transition matrix of the process N_t is simply:

$$Q = \begin{bmatrix} -\lambda(0) & \lambda(0) & 0 & 0 & \cdots \\ \mu(1) & -\lambda(1)-\mu(1) & \lambda(1) & 0 & \cdots \\ 0 & \mu(2) & -\lambda(2)-\mu(2) & \lambda(2) & \cdots \\ & \cdots\cdots\cdots\cdots\cdots\cdots\cdots \\ & \cdots\cdots\cdots\cdots\cdots\cdots\cdots \end{bmatrix}$$

Having defined the transition matrix of the process, we can determine the stationary solution and its existence by considering the solution $v = (v_0, v_1 \ldots)$ of the system $\begin{cases} vQ = 0 \\ v1 = 1. \end{cases}$

Again:

$$\begin{cases} \lambda(0)v_1 = \lambda(0)v_0, \\ \lambda(0)v_0 + \mu(2)v_2 = (\lambda(1) + \mu(1))v_1, \\ \lambda(1)v_1 + \mu(3)v_3 = (\lambda(2) + \mu(2))v_2, \\ \cdots\cdots\cdots\cdots\cdots\cdots\cdots\cdots \\ v_0 + v_1 + v_2 + \cdots = 1. \end{cases}$$

Evidently we obtain:

$$v_i = \frac{\lambda(0)\lambda(1)\ldots\lambda(i-1)}{\mu(1)\mu(2)\ldots\mu(i)} v_0.$$

A necessary and sufficient condition for the existence of a solution is:

$$v_0 \neq 0 \quad \text{and} \quad v_0 + v_1 + \cdots + v_n + \cdots = 1,$$

that is:

$$v_0\left(1 + \sum_{n=0}^{\infty} \frac{\lambda(0)\ldots\lambda(n)}{\mu(1)\ldots\mu(n+1)}\right) = 1.$$

The ergodicity condition of the Markov process becomes:

$$1 + \sum_{n=0}^{\infty} \frac{\lambda(0)\ldots\lambda(n)}{\mu(1)\ldots\mu(n+1)} < \infty.$$

If this condition is satisfied, the distribution limit is simply obtained from the system equations:

$$p(n) = p(0)\frac{\lambda(0)\ldots\lambda(n-1)}{\mu(1)\ldots\mu(n)}$$

and

$$p(0) = \left[1 + \sum_{n=1}^{\infty} \frac{\lambda(0)\ldots\lambda(n-1)}{\mu(1)\ldots\mu(n)}\right]^{-1}.$$

The results of Chapter 1 for the M/M/1 queue are obtained again by putting $\lambda(0) = \lambda(1) = \cdots = \lambda$ and $\mu(1) = \cdots = \mu$. The M/M/1 queue is also a birth and death process in which the birth and death rates do not depend on the system state.

Following the example which we have just treated, several other types of queue have solutions which are practically evident. We shall describe several in the following sections.

3.1.2. M/M/C queue

We now consider a queue with C servers. The arrivals form a Poisson process with parameter λ and the service times are exponentially distributed with rate μ. The service time is independent of the server. If there are n customers in the system with $n \leqslant C$, all the customers are being served. If $n > C$, there is at least one customer waiting in the queue.

The number of customers in the system at time t, denoted by N_t, is a birth and death process. The birth rates are $\lambda(i) = \lambda$ for all i and

$$\mu(1) = \mu, \mu(2) = 2\mu, \ldots, \mu(C) = C\mu, \mu(C+k) = C\mu, k \geqslant 0.$$

The ergodicity condition of the Markov process is given by:

$$1 + \sum_{n=0}^{\infty} \frac{\lambda^n}{\mu(1)\ldots\mu(C)[\mu(C)]^{n-1}} < \infty,$$

which is realized if and only if

$$\rho = \frac{\lambda}{C\mu} < 1.$$

The stationary solution is:

$$p(n) = \begin{cases} p(0)\dfrac{\lambda^n}{n!\,\mu^n} & \text{if } n \leqslant C, \\[2ex] p(0)\dfrac{\lambda^n}{C!\,C^{n-C}\mu^n} & \text{if } n > C, \end{cases}$$

$p(0)$ is obtained by normalization.

The case where the servers are not identical can be treated in the same way by birth and death processes.

3.1.3. M/M/1/m queue

We again have an M/M/1 queue in which the system can accept only m customers to total y including the one who is in service. A new customer who arrives at the full system is not accepted and is definitely rejected. The number of customers N_t in the system at time t is again a birth and death process in which the rates are given by:

$$\begin{cases} \lambda(0) = \lambda(1) \cdots = \lambda(m-1) = \lambda, \\ \lambda(m) = \lambda(m+1) = \cdots = 0, \\ \mu(1) = \mu(2) = \cdots = \mu(n) = \cdots = \mu. \end{cases}$$

In the Markov process, the series $\{0, 1, \ldots, m\}$ is irreducible and recurrent. States $m+1$, $m+2, \ldots$ are transitory. The stability condition is always satisfied. The M/M/1/m queue is always stable since customers are lost if the limit determined by the total capacity of the system is exceeded. The stationary solution is

$$p(n) = p(0)\left(\frac{\lambda}{\mu}\right)^n \quad \text{if } n \leqslant m,$$

$$p(n) = 0 \qquad\qquad \text{if } n > m,$$

which gives

$$p(n) = p^n \frac{1-\rho}{1-\rho^{m+1}} \quad \text{where} \quad \rho = \frac{\lambda}{\mu}.$$

3.1.4. M/M/∞ queue

In this section we consider an M/M/. queue having a sufficient number of servers so that a customer entering the queue always finds at least one free. Customers arrive at the input of the queue (consisting entirely of servers) following a Poisson process with parameter λ and the service times are exponentially distributed with rate μ.

The process N_t is always a birth and death process with rates:

$$\lambda(i) = \lambda \quad \text{for all } i,$$
$$\mu(i) = i\mu \quad \text{for all } i.$$

The ergodicity condition for the Markov process N_t is always satisfied and the stationary solution is only the limiting case of the $M/M/s$ queue treated in Section 1.3.2, that is:

$$p(n) = p(0)\frac{\lambda^n}{n!\,\mu^n} = p(0)\frac{\rho^n}{n!}$$

where $p(0)$ is the normalization constant.

We obtain

$$p(0) = \left[\sum_{n=0}^{\infty}\frac{\rho^n}{n!}\right]^{-1} = e^{-\rho},$$

which gives the solution:

$$p(n) = e^{-\rho}\frac{\rho^n}{n!}, \quad n = 0,1,2,\ldots$$

This is a Poisson distribution. It can be shown that the same solution is obtained for the $M/G/\infty$ queue. That is a queue having a sufficiently large number of servers and a Poisson arrival process has a solution which depends only on the mean of the service time distribution.

Since the stationary solution is a Poisson distribution, the mean number of customers in the queue in the stationary state becomes:

$$E[N] = \rho = \text{mean number of servers occupied.}$$

3.1.5. M/M/m/m queue

The system can accept only m customers but there are only m servers. This is a system with loss: a customer arriving when all the servers are occupied is definitely rejected. The process N_t is a birth and death process with

$$\mu(k) = k\mu \quad \text{if } k = 1,2,\ldots,m,$$
$$\lambda(k) = \lambda \quad \text{if } k < m \quad \text{and} \quad \lambda(k) = 0 \quad \text{if } k \geq m.$$

The stability condition is always satisfied and the stationary solution is given by:

$$p(n) = \begin{cases} p(0)\dfrac{\lambda^n}{n!\mu^n} & \text{if } n \leq m, \\ 0 & \text{otherwise} \end{cases}$$

and

$$p(0) = \left[\sum_{k=0}^{m} \left(\frac{\lambda}{\mu} \right)^k \frac{1}{k!} \right]^{-1},$$

which leads to:

$$p(n) = \frac{\rho^n/n!}{1 + \rho + \rho^2/2! + \cdots + \rho^n/n!}.$$

This particular system has been much used, particularly in telephony, and the stationary solution is called Erlang's formula. It was demonstrated by Erlang in 1917. It must be added that, if the service time is distributed with a general distribution, the solution is exactly the same. That is the $M/G/m/m$ queue has a solution which depends only on the mean service time.

3.1.6. M/M/C/m/K queue

We are going to study here the most general Markovian queue possible: the number of servers C is less than the number of customers m which the system can contain. The quantity K represents the total number of customers who can make use of the system; we suppose this to be greater than the capacity of the station. Customers arrive from the exterior following a Poisson process with parameter $\lambda(K - n)$ where n is the number of customers already present in the system. This is again a Markovian birth and death process with rates:

$$\lambda(n) = \begin{cases} (K - n)\lambda & \text{if } 0 \leqslant n \leqslant m - 1, \\ 0 & \text{otherwise,} \end{cases}$$

$$\mu(n) = \begin{cases} n\mu & \text{if } 0 \leqslant n \leqslant C, \\ C\mu & \text{if } n > C. \end{cases}$$

The ergodicity condition of the birth and death process is always satisfied. The stationary solution is easily obtained:

$$p(n) = p(0) \prod_{i=0}^{n-1} \frac{\lambda(K - i)}{(i + 1)\mu} \quad \text{if } 0 \leqslant n \leqslant C - 1,$$

$$= p(0) \binom{K}{n} \rho^n \quad \text{with } \rho = \frac{\lambda}{\mu},$$

$$p(n) = p(0) \prod_{i=0}^{C-1} \frac{\lambda(K - i)}{(i + 1)\lambda} \prod_{i=C}^{n-1} \frac{\lambda(K - i)}{C\mu},$$

$$= p(0) \binom{K}{n} \frac{n!}{C!} C^{C-n} \rho^n \quad \text{if } C \leqslant n \leqslant m,$$

$$p(n) = 0 \quad \text{if } n > m.$$

$p(0)$ is obtained by normalization of the preceding stationary solution. Its calculation leads to a complex and little used formula; so it is convenient to obtain it by a numerical method.

In the particular case where $m = C$, that is a case with rejection but no waiting, we obtain the following solution:

$$p(n) = \frac{\binom{K}{n}\rho^n}{\sum_{i=0}^{m}\binom{K}{i}\rho^i}$$

which is known as Engset's formula.

This queue has been much used in some important particular cases; firstly, that of the M/M/1/K/K queue which is the model of a system containing K machines which break down after an exponential lifetime with rate λ, and in which there is a single repairer needing an exponentially distributed repair time with rate μ. The diagram for this system is given in Figure 3.1.

In this particular case the solution becomes:

$$p(n) = p(0)\frac{K!}{(K-n)!}\rho^n, \quad n = 0, 1, \ldots, K$$

and

$$p(0) = \left[\sum_{n=0}^{K}\frac{K!}{(K-n)!}\rho^n\right]^{-1}.$$

In particular, use of the server, or the rate of occupation of the repairer, can easily be calculated:

$$\rho = 1 - p(0) = 1 - \frac{v^K/K!}{1 + v + \frac{v^2}{2!} + \cdots + \frac{vK}{K!}},$$

by putting $v = 1/\rho$.

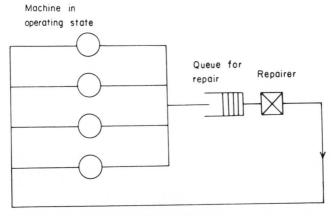

Figure 3.1 M/M/1/K/K queue

A more general model is obtained by increasing the number of repairers. This leads to a M/M/s/K/K system whose solution is the following:

$$p(n) = \begin{cases} p(0)\dbinom{K}{n}\rho^n & \text{if } n \leqslant C, \\[2ex] p(0)\dbinom{K}{n}\dfrac{n!}{C!\,C^{n-C}}\rho^n & \text{if } n > C, \end{cases}$$

and $p(0)$ is obtained by normalization:

$$p(0) = \left[\sum_{n=0}^{C} \binom{K}{n}\rho^n + \sum_{n=C+1}^{K} \binom{K}{n}\frac{n!}{C!\,C^{n-C}}\rho^n \right]^{-1}.$$

3.1.7. M/M/C queue with possible departure

The queue which we wish to study is the classic M/M/C queue in which a customer joining the queue decides whether he will wait for or renounce his service. In the case where he renounces, he definitely leaves. We denote by r_i the probability that the customer stays in the queue if he finds i customers in front of him. It must be noted that if the customer stays in the queue, he will stay there until the end of his service. The process N_t is again Markovian, it is a birth and death process with rates:

$$\lambda(0) = \lambda r_0, \ \lambda(1) = \lambda r_1, \ldots, \lambda(n) = \lambda r_n, \ldots,$$
$$\mu(1) = \mu, \mu(2) = 2\mu, \ldots, \mu(C) = C\mu,$$
$$\mu(C+1) = \mu(C+2) = \cdots = C\mu.$$

The solution for the equilibrium state is given by:

$$p(n) = p(0)\frac{\lambda^n \prod\limits_{i=0}^{n-1} r_i}{n!\,\mu^n} \qquad \text{if } n \leqslant C$$

$$p(n) = p(0)\frac{\lambda^n \prod\limits_{i=0}^{n-1} r_i}{C!\,C^{n-C}\mu^n} \qquad \text{if } n > C$$

and

$$p(0) = \left[1 + \sum_{n=1}^{C} \frac{\lambda^n}{n!\,\mu^n} \prod_{i=0}^{n-1} r_i + \sum_{n=C+1}^{\infty} \frac{\lambda^n}{C!\,C^{n-C}\mu^n} \prod_{i=0}^{n-1} r_i \right]^{-1}$$

The stationary solution exists if $p(0) > 0$, which can be expressed by:

$$\limsup_{i \to \infty} \lambda r_i < C\mu.$$

3.2 - DIFFUSION PROCESS APPROXIMATION

3.2.1. Introduction

Consider a GI/GI/1 queue with first come, first served service discipline. Let $N(t)$ be the number of customers at this station at time t. We assume that the time intervals between arrivals and the service durations are independent and identically distributed random variables characterized by their mean and variance: $1/\lambda = $ mean of the interarrivals, $1/\mu = $ mean of the service times, $\mathrm{Var}\,a = $ variance of the interarrivals, $\mathrm{Var}\,s = $ variance of the service times. We shall put $Ka = \mathrm{Var}\,a\lambda^2$ and $Ks = \mathrm{Var}\,s\mu^2$. The quantities Ka and Ks are the squares of the variation coefficients (SVC) of the interarrivals and service times respectively.

Again let $A(t)$ and $D(t)$ be the number of customers entering and leaving the queue respectively between the instants 0 and t. The number of customers in the queue at time t is given by $N(t) = A(t) - D(t) + N_0$ where N_0 is the number of customers present in the queue initially.

When the queue is not empty, the process $N(t)$ has the following properties: if $t > s$, $N(t) - N(s)$ is independent of $N(s)$; if $t - s$ is sufficiently large, $N(t) - N(s)$ tends to a normal law with derivative $\beta(t - s)$ and variance $\alpha(t - s)$. In fact, this difference between two processes tending to a normal law is a normal distribution process if the two processes are independent. That is one of the difficulties of the problem, since, in general, the departure process depends on the arrival process. However, if the number of customers in the system is not zero, the independence criteria are justified. It is for this reason that the diffusion process approximation has already been applied to these systems when saturated or heavily loaded.

The values of α and β are obtained in the following manner:

$$\beta(t) = \lim_{\Delta t \to 0} \frac{E\{N(t + \Delta t) - N(t)\}}{\Delta t},$$

$$\alpha(t) = \lim_{\Delta t \to 0} \frac{E\{(N(t + \Delta t) - N(t))^2\}}{\Delta t} - \frac{(E\{N(t + \Delta t) - N(t)\})^2}{\Delta t}$$

In the case which interests us (the GI/GI/1 queue), by applying the limit theorem of renewal processes, we have in the equilibrium state ($t \to \infty$):

$$\beta = \lambda - \mu, \quad \alpha = \lambda Ka + \mu Ks.$$

We can state the following theorem:

Theorem

If (i) $N(t) - N(s)$ is independent of $N(s)$ for $s < t$,

(ii) $N(t) - N(s)$ is distributed according to a normal (Gaussian) law with mean $\beta(t - s)$ and variance $\alpha(t - s)$,

(iii) $N(0) = N_0$,

then:

$$N(t) = N_0 + \beta t + \alpha b(t)$$

where $b(t)$ is normalized Brownian motion, that is:

$b(0) = 0$

$b(t) - b(s)$ follows the law $\mathcal{N}(0, t - s)$, a Gaussian variable with mean 0 and variance $(t - s)$,

$b(t) - b(s)$ is independent of $b(s)$.

Proof: From property (ii), we have:

$$P\{N(t) - N(s) < y\} = \int_{-\infty}^{y} \frac{1}{\sqrt{2\pi(t - s)\alpha}} \exp\left[-\frac{(u - \beta(t - s))^2}{2\alpha(t - s)} \right] du \quad (3.2.1)$$

Putting

$$u' = \frac{u - \beta(t - s)}{\sqrt{\alpha}},$$

$$du' = \frac{du}{\sqrt{\alpha}},$$

equality (3.2.1) is then equivalent to:

$$\int_{-\infty}^{(y - \beta(t - s))/\sqrt{\alpha}} \frac{1}{\sqrt{2\pi(t - s)}} \exp\left[-\frac{u'^2}{2(t - s)} \right] du' = P\left[b(t) - b(u) < \frac{y - \beta(t - s)}{\sqrt{\alpha}} \right]$$

$$(3.2.2)$$

Due to property (i) this implies that:

$$N(t) = n_0 + \beta t + \sqrt{\alpha} b(t).$$

Also property (iii) determines the constant $n_0 = N_0$ and hence:

$$N(t) = N_0 + \beta t + \sqrt{\alpha} b(t).$$

Theorem

The probability density of the process $N(t)$:

$$f(N_0, x, t) = \frac{\partial}{\partial x} P[N(t) \leqslant x | N(0) = N_0]$$

is given by:

$$f(N_0, x, t) = \frac{1}{\sqrt{2\pi t \alpha}} \exp\{ -(x - N_0 - \beta t)^2 / 2\alpha t \}.$$

Proof

$$P[N(t) \leqslant x | N(0) = N_0] = P[N_0 + \beta t + \sqrt{\alpha} b(t) \leqslant x],$$

$$= P\left[b(t) \leqslant \frac{x - \beta t - N_0}{\sqrt{\alpha}} \right], \qquad (3.2.3)$$

$$= P\left[b(t) - b(0) \leqslant \frac{x - \beta t - N_0}{\sqrt{\alpha}} \right]$$

$$P[N(t) \leqslant x | N(0) = N_0] = \int_{-\infty}^{(x - \beta t - N_0)\sqrt{\alpha}} \frac{1}{\sqrt{2t}} \exp\left(-\frac{y^2}{2t^2} \right) dy. \qquad (3.2.4)$$

By putting:

$$y = (u - \beta t - N_0)/\sqrt{\alpha},$$

$$dy = \frac{du}{\sqrt{\alpha}},$$

then

$$y = (x - N_0 - \beta t)/\sqrt{\alpha} \Rightarrow u = x.$$

Equation (3.2.4) is equivalent to:

$$P[N(t) \leqslant x | N(0) = N_0] = \int_{-\infty}^{x} \frac{1}{\sqrt{2\Pi t \alpha}} \exp\left[\frac{-(u - N_0 - \beta t)^2}{2t^2 \alpha} \right] du,$$

from which one deduces:

$$\frac{\partial}{\partial x} P[N(t) \leqslant x | N(0) = N_0] = \frac{1}{\sqrt{2\Pi t \alpha}} \exp\left[\frac{-(x - N_0 - \beta t)^2}{2\alpha t^2} \right].$$

Theorem

The expression

$$f(N_0, x, t) = \frac{1}{\sqrt{2\Pi t \alpha}} \exp\left[\frac{-(x - N_0 - \beta t)^2}{2\alpha t^2} \right]$$

satisfies the diffusion equation or the Fokker–Planck equation:

$$\frac{\partial}{\partial t} f(N_0, x, t) = \frac{1}{2} \alpha \frac{\partial^2}{\partial x^2} f(N_0, x, t) - \beta \frac{\partial}{\partial x} f(N_0, x, t). \qquad (3.2.5)$$

Proof The reader will be able to establish the proof of this theorem for himself; for this it is sufficient to verify equation (3.2.5).

We have shown that the process determining the length of the queue has a probability density which confirms the diffusion equation.

Comments A slightly different presentation can be given starting with a difference equation for $N(t)$:

$$dN(t) = dA(t) - dD(t).$$

In this case, $dN(t)$ satisfies the equation:

$$dN(t) = \beta dt + \sqrt{\alpha} db(t)$$

which is the stochastic differential in the sense of Ito of the equation:

$$N(t) = N_0 + \beta t + \sqrt{\alpha} b(t).$$

The probability distribution of $N(t)$ is therefore Brownian motion with derivative β and variance α. However, the condition $N(t) \geq 0$ has not been taken into account. In fact the number of customers in the system must be positive or zero.

Therefore it is necessary to introduce a barrier at the origin. Several types of barrier can be used. We shall study two: the reflecting barrier and the absorbing barrier with which we associate an instantaneous return process at point 1. In the following sections we shall suppose that an equilibrium state exists which makes the function $f(N_0, x, t)$ independent of N_0 and t.

3.2.2. Reflecting barrier

This type of barrier is that which one thinks of first. When the process reaches the origin it is instantaneously reflected to the positive part of the axis. The advantage of this type of barrier arises from the fact that the assumption $N(t) > 0$ is always true except for a negligible series in the sense of a Lebesgue measure.

From (3.2.5) we obtain:

$$\int_{0^+}^{\infty} \frac{\partial f(x,t)}{\partial t} dx = \int_{0^+}^{\infty} \left[-\beta \frac{\partial f(x,t)}{\partial x} + \frac{\alpha}{2} \frac{\partial^2 f(x,t)}{\partial x^2} \right] dx \qquad (3.2.6)$$

The term on the left cancels out since $f(x,t)$ is a probability distribution. As we have assumed that an equilibrium state exists, equation (3.2.6) becomes:

$$\beta f(0^+) = \frac{\alpha}{2} \left[\frac{\partial f}{\partial x} \right]_{x=0_+} \qquad (3.2.7)$$

The condition $\int_0^{\infty} f(x) dx = 1$ implies the existence of a unique solution:

$$f(x) = -\gamma e^{\gamma x} \quad \text{if } x \geq 0,$$
$$f(x) = 0 \qquad \text{if } x < 0$$

where

$$\gamma = \frac{2\beta}{\alpha} = \frac{-2(1-\rho)}{\rho K a + K s} \quad \text{if } \gamma < 0 \left(\rho = \frac{\lambda}{\mu} < 1 \right).$$

Before considering possible discretizations, let us consider a second type of barrier. In fact the technique which we have just described implies that the mass at the origin is zero which is true only if $\rho = 1$. To increase the range of the diffusion approximation in a realistic manner, the following approach is desirable.

3.2.3. Absorbing barrier and instantaneous return

Consider an absorbing barrier at the point $x = 0$, which introduces a mass at this point into the probability distribution. At the end of the absorption the process jumps to point 1. The mass at point 0 corresponds to the probability that the system is empty and the jump to point 1 corresponds to the arrival of a customer in the queue.

In this case, the diffusion equation must be modified and extended to take account of the absorption and instantaneous return to point 1.

In the general case, let ℓ be the point of instantaneous return, $\delta(x = \ell)$ the Dirac function at the point ℓ, Λ^{-1} the mean time of the period of inactivity, and $R(t)$ the probability that the process reaches the boundary at time t. The diffusion equations become:

$$\frac{\partial f(x,t)}{\partial t} + \beta \frac{\partial f(x,t)}{\partial x} - \frac{\alpha}{2}\frac{\partial^2 f(x,t)}{\partial x^2} = \Lambda R(t)\delta(x = \ell). \qquad (3.2.8)$$

On the other hand the probability density must satisfy a conservation equation for waves on the boundary:

$$\frac{dR(t)}{dt} = -\Lambda R(t) + \lim_{x \to 0}\left(-\beta f(x,t) + \frac{1}{2}\frac{\partial f(x,t)}{\partial x}\right). \qquad (3.2.9)$$

Let us put $R = \lim_{t \to \infty} R(t)$. The function $f(x) = \lim_{t \to \infty} f(x,t)$ must satisfy equations (3.2.5) and (3.2.6) as t tends to infinity and also the normalization condition:

$$R + \int_{0^+}^{\infty} f(x)\,dx = 1.$$

The solution to these equations is unique and of the form:

$$f(x) = \begin{cases} \rho(e^{-\gamma} - 1)e^{\gamma x} & \text{if } x \geq 1, \\ \rho(1 - e^{\gamma x}) & \text{if } 0 \leq x \leq 1. \end{cases} \qquad (3.2.10)$$

where

$$\gamma = \frac{2\beta}{\alpha} = \frac{-2(1-\rho)}{\rho K a + K s} \quad \text{and} \quad R = 1 - \frac{\Lambda}{\Lambda + \mu - \lambda}.$$

We know that $R = 1 - \rho$ which implies $\Lambda = \lambda$.

This last equality is exact if the arrivals form a Poisson process; in other cases it is merely a good approximation. In the following we shall take $R = 1 - \rho$.

3.2.4. Discretization of the continuous process

Since we have replaced a jump process by a continuous process, the equilibrium state is described in the form of a continuous probability distribution. We can return to a discrete distribution in several ways. We shall define three. Let $p(k)$ be the probability that there are k customers in the queue.

1. Integrate the continuous function $f(x)$ over the interval $[k, k+1]$:

$$p(k) = \int_k^{k+1} f(x)\,dx.$$

2. Integrate $f(x)$ between $k-1$ and k:

$$p(k) = \int_{k-1}^k f(x)\,dx, \quad k = 1, \ldots .$$

3. Put $p(k) = f(k)$.

In the three cases, it is necessary to have $\sum_{k=0}^{\infty} p(k) = 1$.

With a reflecting barrier, it is preferable to use one of the first two methods since the normalization condition is automatically satisfied.

In the case of the continuous function $f(x) = -\gamma e^{\gamma x}$ and using the first discretization procedure, we obtain:

$$p(k) = \int_k^{k+1} f(x)\,dx = (1 - \hat{\rho})\hat{\rho}^k, \quad k = 0, 1, 2\ldots, \tag{3.2.11}$$

where

$$\hat{\rho} = e^{\gamma} = \exp(2\beta/\alpha) = \exp(2(\lambda - \mu)/(\lambda K a + \mu K s)).$$

This form of distribution is well known, it is of a type identical to that of the M/M/1 queue: $(1 - \rho)\rho^k$, $k = 0, 1, 2\ldots$. As $\hat{\rho}$ is a very good approximation to ρ when $Ka = Ks = 1$, in this case we have a very good approximation to $p(k)$.

However, because of the reflecting barrier, the value of $p(0)$ alters rapidly when the squares of the variation coefficients (SVC) move away from 1. Also it is recommended to adjust the distribution in the following manner:

$$p(k) = \begin{cases} 1 - \rho & \text{if } k = 0, \\ \rho(1 - \hat{\rho})\hat{\rho}^{k-1} & \text{if } k > 1. \end{cases} \tag{3.2.12}$$

With this approximation the error in the mean number of customers in the system is very small for the M/G/1 queue if the SVC of the service time Ks is close to 1. This error increases as Ks moves away from 1. However, the relative error of this value becomes zero as ρ tends to 1.

The error relative to the length of the M/G/1 queue is represented in Figure 3.2 as a function of the value of the traffic rate ρ, for different values of the SVC of the service time.

In the case where an absorbing barrier with instantaneous return to point 1 is chosen, type 1 discretization cannot be used. In fact this technique alone

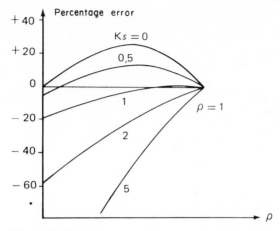

Figure 3.2 Relative error of the mean length of the M/G/1 queue

allows the value of $p(0)$ to be calculated: it is obtained directly from the mass on the barrier.

The second discretization procedure, $p(k) = \int_{k-1}^{k} f(x) \, dx$, gives the following values:

$$\begin{cases} p(0) = 1 - \rho \\ p(1) = \dfrac{\rho(\rho\,Ka + Ks)}{2(1-\rho)}(e^{\gamma} - 1 - \lambda), \\ p(k) = \dfrac{\rho(\rho\,Ka + Ks)}{2(1-\rho)}(e^{k\gamma}(1 - e^{-\gamma})), \quad k \geqslant 2. \end{cases}$$

This relatively complicated probability distribution gives an interesting result for the mean number of customers $E[n_2]$:

$$E[n_2] = \sum_{1}^{\infty} kp(k) = \rho\left(1 + \frac{\rho\,Ka + Ks}{2(1-\rho)}\right).$$

It is interesting to note that $E[n_2]$ has a similar form to the Pollaczek and Khintchine formula for the M/G/1 queue:

$$E[n] = \rho\left(1 + \frac{\rho(1 + Ks)}{2(1-\rho)}\right).$$

The absolute error is $E[n_2] - E[n] = (1/2)\rho\,Ks$ by putting $Ka = 1$.

The third procedure defined by $p(k) = f(k)$ gives the adjusted solution again exactly:

$$p(k) = \begin{cases} 1 - \rho & k = 0, \\ \rho(1 - \hat{\rho})\hat{\rho}^{k-1}, & k \geqslant 1 \quad \text{with} \quad \hat{\rho} = e^{\gamma}. \end{cases}$$

This solution has been found again without starting from an already known result. The mean number calculated from this discretization is:

$$E[n_1] = \frac{\rho}{1 - \hat{\rho}} \quad \text{with} \quad \hat{\rho} = \exp\left[\frac{2(\lambda - \mu)}{\lambda Ka + \mu Ks}\right].$$

Another discretization procedure for continuous functions will now be described. It extends the application area of diffusion processes but also introduces numerical calculations. We briefly describe this procedure: the probability of having i customers in the system is given by

$$P(i) = \int_{b_{i-1}}^{b_i} f(x)\,dx \quad \text{for} \quad i = 1,\ldots$$

where b_1 is determined by the following equation:

$$\int_0^{b_1} x f(x)\,dx = \int_0^{b_1} f(x)\,dx \quad \text{and} \quad b_0 = 0;$$

and

$$b_j = b_{j-1} + 1.$$

A slightly different method can also be proposed, by taking account of the fact that $\Delta D(t)$ has an SVC equal to Ks during a proportion ρ of the time. This introduces a factor ρ into the calculation of the variance of the normal law which approximates the probability distribution of $\Delta N(t)$. In this case we have:

$$\beta = \lambda - \mu \quad \text{and} \quad \alpha' = \lambda Ka + \mu Ks\rho.$$

By putting $\gamma' = 2\beta/\alpha' = (2\lambda - \mu)/\lambda(Ka + Ks)$, we obtain a solution of the diffusion equation identical to that found by Gelenbe [11] by replacing only α by α'.

With the third procedure ($p(k) = f(k)$), the distribution in the equilibrium state is written:

$$\begin{cases} p(0) = 1 - \rho, \\ p(k) = (1 - \hat{\rho}')\hat{\rho}'^{k-1} & \text{if} \quad s \geqslant 1, \end{cases}$$

with

$$\hat{\rho}' = e^{\gamma'} = \exp\left[\frac{2(\lambda - \mu)}{\lambda(Ka + Ks)}\right].$$

We shall call $E[n_3]$ the mean number of customers in the system calculated from this discretization. We have:

$$E[n_3] = \sum_{k=1}^{\infty} kP(k) = \frac{\rho}{1 - \hat{\rho}'}.$$

Also the probability distribution of the mean number of customers in the queue has a form analogous to the results obtained for the M/M/1 queue since, as we have already said, $\hat{\rho}'$ plays the same role as ρ for diffusions.

Adopting a type 2 discretization this gives

$$p(k) = \int_{k-1}^{k} f(x)\,dx,$$

we obtain the following solution:

$$\begin{cases} p(0) = 1 - \rho \\[2mm] p(1) = \dfrac{\rho(Ks + Ka)}{2(1 - \rho)}(e^{\lambda'} - 1 - \lambda') \\[3mm] p(k) = \dfrac{\rho(Ks + Ka)}{2(1 - \rho)}e^{k\lambda'}(1 - e^{-\lambda'}), \quad k \geqslant 2. \end{cases}$$

Therefore the mean length of the queue is:

$$E[n_4] = \int_0^\infty x f(x)\,dx = \rho\left[1 + \frac{\rho(Ka + Ks)}{2(1 - \rho)}\right].$$

In this case we again find the approximate formula proposed by Kingman (cf. Section 1.6) for calculating the mean length of a GI/GI/1 queue. In particular, in the case where $Ka = 1$, the formula is identical to that of Pollaczek and Khintchine.

To give an idea of the accuracy of the results for the different cases which we have just described, we have presented, in Table 3.1, the values of the mean number of customers in a type E_2/M/1 queue (Erlang-2 distribution of intervals between arrivals). The exact calculation of the mean number of customers in the system is made from the following formula:

$$E[n] = \frac{\rho}{1 - \sigma}$$

where σ is the unique root of $\sigma = A^*(\mu - \mu\sigma)$ and A^* is the Laplace–Stieltjes transform of the distribution of the intervals between arrivals. Table 3.1 contains several values of traffic rate ρ.

In Tables 3.2 and 3.3, we compare the four diffusion approximation formulae in a systematic manner for two values of traffic rate $\rho = 0.8$ and $\rho = 0.5$. The

Table 3.1

ρ	$E[n]$ (exact results)	$E[n_1]$	$E[n_2]$	$E[n_3]$	$E[n_4]$
0.95	14.331	14.492	14.962	14.144	14.487
0.90	6.829	6.985	7.425	6.194	6.974
0.85	4.327	4.887	4.887	4.057	4.463
0.80	3.075	3.219	3.600	2.800	3.200
0.75	2.323	2.460	2.813	2.905	2.438
0.70	1.820	1.95	2.275	1.724	1.925

Table 3.2 - **Mean number of customers for $\rho = 0.5$.**

Ka	Ks	E[n]	E[n_1]	E[n_2]	E[n_3]	E[n_4]
.0	.0	.80	.80	.80	.80	.80
.0	.5	1.37	1.45	1.80	1.27	1.60
.0	1.0	2.15	2.43	2.80	2.03	2.40
.0	1.5	2.95	3.42	3.80	2.82	3.20
.0	2.0	3.74	4.41	4.80	3.62	4.00
.0	2.5	4.54	5.41	5.80	4.41	4.80
.0	3.0	5.34	6.41	6.80	5.21	5.60
.5	.0	1.54	1.27	1.60	1.27	1.60
.5	.5	2.33	2.23	2.60	2.03	2.40
.5	1.0	3.13	3.22	3.60	2.82	3.20
.5	1.5	3.93	4.21	4.60	3.62	4.00
.5	2.0	4.73	5.21	5.60	4.41	4.80
.5	2.5	5.53	6.21	6.60	5.21	5.60
.5	3.0	6.33	7.21	7.60	6.01	6.40
1.0	.0	2.40	2.03	2.40	2.03	2.40
1.0	.5	3.20	3.02	3.40	2.82	3.20
1.0	1.0	4.00	4.01	4.40	3.62	4.00
1.0	1.5	4.80	5.01	5.40	4.41	4.80
1.0	2.0	5.60	6.01	6.40	5.21	5.60
1.0	2.5	6.40	7.01	7.40	6.01	6.40
1.0	3.0	7.20	8.01	8.40	6.81	7.20
1.5	.0	3.05	2.82	3.20	2.82	3.20
1.5	.5	3.91	3.82	4.20	3.62	4.00
1.5	1.0	4.73	4.81	5.20	4.41	4.80
1.5	1.5	5.54	5.81	6.20	5.21	5.60
1.5	2.0	6.34	6.81	7.20	6.01	6.40
1.5	2.5	7.14	7.81	8.20	6.81	7.20
1.5	3.0	7.95	8.81	9.20	7.61	8.00
2.0	.0	3.70	3.62	4.00	3.62	4.00
2.0	.5	4.60	4.61	5.00	4.41	4.80
2.0	1.0	5.44	5.61	6.00	5.21	5.60
2.0	1.5	6.26	6.61	7.00	6.01	6.40
2.0	2.0	7.07	7.61	8.00	6.81	7.20
2.0	2.5	7.88	8.61	9.00	7.61	8.00
2.0	3.0	8.69	9.61	10.00	8.41	8.80
2.5	.0	4.35	4.41	4.80	4.41	4.80
2.5	.5	5.29	5.41	5.80	5.21	5.60
2.5	1.0	6.15	6.41	6.80	6.01	6.40
2.5	1.5	6.98	7.41	7.80	6.81	7.20
2.5	2.0	7.80	8.41	8.80	7.61	8.00
2.5	2.5	8.61	9.41	9.80	8.41	8.80
2.5	3.0	9.42	10.41	10.80	9.21	9.60
3.0	.0	5.00	5.21	5.60	5.21	5.60
3.0	.5	5.97	6.21	6.60	6.01	6.40
3.0	1.0	6.84	7.21	7.60	6.81	7.20
3.0	1.5	7.69	8.21	8.60	7.61	8.00
3.0	2.0	8.51	9.21	9.60	8.41	8.80
3.0	2.5	9.33	10.21	10.60	9.21	9.60
3.0	3.0	10.15	11.20	11.60	10.01	10.40

Table 3.3 - **Mean number of customers for $\rho = 0.8$.**

Ka	Ks	E[n]	E[n_1]	E[n_2]	E[n_3]	E[n_4]
.0	.0	.50	.50	.50	.50	.50
.0	.5	.53	.58	.75	.51	.63
.0	1.0	.63	.79	1.00	.58	.75
.0	1.5	.74	1.03	1.25	.68	.83
.0	2.0	.86	1.27	1.50	.79	1.00
.0	2.5	.98	1.52	1.75	.91	1.13
.0	3.0	1.10	1.76	2.00	1.03	1.25
.5	.0	.59	.51	.63	.51	.63
.5	.5	.71	.68	.88	.58	.75
.5	1.0	.84	.91	1.13	.68	.88
.5	1.5	.96	1.15	1.38	.79	1.00
.5	2.0	1.08	1.39	1.63	.91	1.13
.5	2.5	1.21	1.64	1.88	1.03	1.25
.5	3.0	1.33	1.89	2.13	1.15	1.38
1.0	.0	.75	.58	.75	.58	.75
1.0	.5	.88	.79	1.00	.68	.88
1.0	1.0	1.00	1.03	1.25	.79	1.00
1.0	1.5	1.13	1.27	1.50	.91	1.13
1.0	2.0	1.25	1.52	1.75	1.03	1.25
1.0	2.5	1.38	1.76	2.00	1.15	1.38
1.0	3.0	1.50	2.01	2.25	1.27	1.50
1.5	.0	.82	.68	.88	.68	.88
1.5	.5	.97	.91	1.13	.79	1.00
1.5	1.0	1.10	1.15	1.38	.91	1.13
1.5	1.5	1.23	1.39	1.63	1.03	1.25
1.5	2.0	1.35	1.64	1.88	1.15	1.38
1.5	2.5	1.48	1.89	2.13	1.27	1.50
1.5	3.0	1.60	2.14	2.38	1.39	1.63
2.0	.0	.89	.79	1.00	.79	1.00
2.0	.5	1.05	1.03	1.25	.91	1.13
2.0	1.0	1.19	1.27	1.50	1.03	1.25
2.0	1.5	1.32	1.52	1.75	1.15	1.38
2.0	2.0	1.45	1.76	2.00	1.27	1.50
2.0	2.5	1.58	2.01	2.25	1.39	1.63
2.0	3.0	1.71	2.26	2.50	1.52	1.75
2.5	.0	.96	.91	1.13	.91	1.13
2.5	.5	1.13	1.15	1.38	1.03	1.25
2.5	1.0	1.28	1.39	1.63	1.15	1.38
2.5	1.5	1.42	1.64	1.88	1.27	1.50
2.5	2.0	1.55	1.89	2.13	1.39	1.63
2.5	2.5	1.68	2.14	2.38	1.52	1.75
2.5	3.0	1.81	2.38	2.63	1.64	1.88
3.0	.0	1.04	1.03	1.25	1.03	1.25
3.0	.5	1.22	1.27	1.50	1.15	1.38
3.0	1.0	1.37	1.52	1.75	1.27	1.50
3.0	1.5	1.51	1.76	2.00	1.39	1.63
3.0	2.0	1.64	2.01	2.25	1.52	1.75
3.0	2.5	1.77	2.26	2.50	1.64	1.88
3.0	3.0	1.90	2.51	2.75	1.76	2.00

SVC of the interarrivals and the service times vary between 0 and 3 in steps of 0.5. These results are compared with those of a simulation or with exact results if they exist (Pollaczek and Khintchine). For squares of variation coefficients greater than 1, we have adopted the hyperexponential distribution with two servers and equal probability of going to one or the other. The service times are chosen to obtain the required square of the variation coefficient.

Comparing Tables 3.1, 3.2 and 3.3, we cannot say that one formula is clearly more adequate than another, except that the values obtained for $E[n_4]$ (which is also Kingman's classic formula) seem to be generally the most correct.

We have also compared simulation results for queues having the same square of variation coefficient, but not the same distribution, with those obtained by diffusion. In the case of hyperexponential interarrivals, the difference between these cases can reach 30%. Introduction of the third moment, by any method, would be desirable in this case.

Before finishing this section, restricted to the particular case of a queue with a single server, another advantage of the diffusion process approximation must be mentioned: it is the possibility of obtaining a solution when the equilibrium state has not been reached. Let x_0 be the initial state. The solution for a transient state contains terms which decrease with time and become zero as $t \to \infty$. These terms show the effect of time on the initial value x_0. The non-stationary solution of several simple models is proposed by Kobayashi [18].

3.3 - PARTICULAR QUEUES

3.3.1 Queue with batch arrivals $M^x/M/1$

We are going to study the $M/M/1$ queue for which the arrivals occur in groups. The number of customers per group is a strictly positive random variable X. We write it:

$$P\{X = x\} = C_x.$$

This system is Markovian since future behaviour depends uniquely on the present situation. We have $C_x = \lambda_x/\lambda$, if λ_x is the rate of arrival of groups containing x customers.

We can write the equilibrium state equations very simply using the usual notation:

$$\begin{cases} (\lambda + \mu)p(n) = \mu p(n + 1) + \lambda \sum_{k=1}^{n} p(n - k)C_k, & n \geqslant 1, \\ \lambda p(0) = \mu p(1). \end{cases} \tag{3.3.1}$$

To resolve this system, we shall use the following two generating functions:

$$P(x) = \sum_{n=0}^{\infty} p(n)x^n$$

and

$$C(x) = \sum_{n=0}^{\infty} C_n x^n.$$

Multiplying equations (3.3.1) by the corresponding x^n and summing the series, we obtain:

$$\lambda \sum_{n=0}^{\infty} p(n)x^n + \mu \sum_{n=1}^{\infty} p(n)x^n = \frac{\mu}{x} \sum_{n=1}^{\infty} p(n)x^n + \lambda \sum_{n=1}^{\infty} \sum_{k=1}^{n} p(n-k)C_k x^n. \qquad (3.3.2)$$

The last term can be simply written:

$$\lambda \sum_{n=1}^{\infty} \sum_{k=1}^{n} p(n-k)C_k x^n = \lambda \sum_{k=1}^{\infty} C_k x^k \sum_{n=k}^{\infty} p(n-k)x^{n-k},$$

$$= \lambda C(x)P(x). \qquad (3.3.3)$$

Equation (3.3.2) becomes:

$$\lambda P(x) + \mu[P(x) - p(0)] = \frac{\mu}{x}[P(x) - p(0)] + \lambda C(x)P(x) \qquad (3.3.4)$$

and we obtain the generating function:

$$P(x) = \frac{\mu p(0)(1-x)}{\mu(1-x) - \lambda x[1-C(x)]} \quad \text{if} \quad |x| < 1.$$

To obtain the value $p(0)$, it is necessary to use the condition $P(1) = 1$. Relation (3.3.4), as x tends to 1, and on noting $\alpha = \lim_{x \to 1}(dC(x)/dx)$, the average number of groups arriving, becomes:

$$1 = \frac{-\mu p(0)}{-\mu + \lambda \alpha},$$

hence

$$p(0) = 1 - \frac{\lambda \alpha}{\mu},$$

$$p(0) = 1 - \rho,$$

by putting:

$$\rho = \frac{\lambda \alpha}{\mu}.$$

The stability condition for the queue is obtained for $p(0) > 0$ that is $\rho < 1$.

Example

Suppose that the arrivals of groups are geometrically distributed and hence follow:

$$C_n = (1-a)a^{n-1}, \quad n \geqslant 1,$$

which gives:

$$P(x) = (1 - \rho) \left[\frac{1}{1 - [a + (1 - a)\rho]x} - \frac{ax}{1 - [a + (1 - a)\rho]x} \right], \quad (3.3.5)$$

we obtain:

$$p(n) = \left[\frac{\partial^n P(x)}{\partial x^n} \right]_{x=0}$$

which gives:

$$p(0) = 1 - \frac{\lambda \alpha}{\mu} = 1 - \rho,$$

$$p(n) = (1 - \rho)[a + (1 - a)\rho]^{n-1}[(1 - a)\rho], \quad n > 0. \qquad (3.3.6)$$

3.3.2. Queue with batch service: $M/M^Y/1$

Customers are always served in 'first come, first served' order. We shall suppose that customers are served in groups of size K, unless the number of customers present in the queue does not allow this size to be reached. In this last case all the customers in the queue are served together. We shall suppose that the distribution of service times does not depend on the size of the group being served and that it is exponential with parameter μ.

The Chapman–Kolmogorov equations at the equilibrium state are easily written:

$$\begin{cases} (\lambda + \mu)p(n + 1) = \lambda p(n) + \mu p(n + K + 1), \\ \lambda p(0) = \mu p(1) + \cdots + \mu p(K - 1) + \mu p(K). \end{cases} \qquad (3.3.7)$$

We can write this system of equations with the help of the operator $\hat{p}$ defined by $\hat{p}(p(n)) = p(n + 1)$:

$$[\mu \hat{P}^{K+1} - (\lambda + \mu)\hat{P} + \lambda]p(n) = 0, \quad n \geq 0.$$

Let us call the roots of this functional equation $r_1, \ldots, r_{K+1}$. Constants C_i exist such that:

$$p(n) = \sum_{i=1}^{K+1} C_i r_i^n.$$

Since $p(n)$ is a probability, $C_i = 0$ if r_i is not a root whose modulus is less than 1. The number of roots whose modulus is less than 1 is obtained from Rouche's theorem: we find that there is only one which we denote by r_0.

Normalizing, it becomes:

$$p(0) = 1 - r_0,$$
$$p(n) = (1 - r_0)r_0^n$$

We immediately obtain:

$$E[N] = \frac{r_0}{1 - r_0},$$

$$E[W] = \frac{r_0}{\lambda(1 - r_0)}.$$

3.4 - QUEUES WITH PRIORITY

So far a single class of customer has been considered. This leads to the fact that all customers are served in an identical manner. In this section we are going to introduce several classes of customer. Numerous examples will be treated by taking account of only two classes of customer. If the queue has a first in, first out service discipline, it can be seen that it is always necessary to know the class number of the customer entering service in order to provide him with the service corresponding to his class. The system state is determined by the number of customers in each class and by the set of class numbers of the customers in the queue. Stationary state determination is particularly complicated because of this. To treat a problem with several classes of customers, it is necessary to define priority rules for the different classes of customer. We shall suppose in this section that class 1 has higher priority than class 2, that class 2 has higher priority than class 3, etc. To define completely the service discipline, it is necessary to determine whether the priority is absolute or not. By absolute priority, we mean that a customer of lower priority is returned to the head of the queue when a customer of higher priority arrives, in the queue, the later arrival starting his service immediately. If the priority is not absolute, a new customer of higher priority waits for the service to be finished before starting his. In the case of absolute priority two new possibilities arise: the customer may resume his service where he was interrupted or he may resume it from the start.

3.4.1. M/M/1 queue with different customer classes and absolute priorities

In systems with absolute priority, customers with a higher priority are not affected in any way by those with a lower priority. Suppose that there are two classes of customer C_1 and C_2 which represent proportions α and $1 - \alpha$ of the customers. Their service times follow an exponential distribution with parameters μ_1 and μ_2 respectively. The arrival process is Poisson with parameter λ_1 for class 1 and λ_2 for class 2, parameters which we shall also denote $\lambda_1 = \lambda\alpha$ and $\lambda_2 = \lambda(1 - \alpha)$.

Given that class 1 is not affected by class 2, for customers in class 1 in the equilibrium state we obtain:

$$p_1(n) = \rho_1^n(1 - \rho_1) \quad \text{or} \quad \rho_1 = \frac{\lambda_1}{\mu_1}.$$

As the service times are exponentially distributed, resuming service where it was interrupted and restarting at the beginning are equivalent.

Let us write the Chapman–Kolmogorov equations in the equilibrium state where $p(n_1, n_2)$ indicates the probability of having n_1 class C_1 customers and n_2 class C_2 customers in the system.

$$(\lambda_1 + \lambda_2 + \mu_1)p(n_1, n_2) = \lambda_1 p(n_1 - 1, n_2) + \mu_1 p(n_1 + 1, n_2)$$
$$+ \lambda_2 p(n_1, n_2 - 1) \quad \text{if} \quad n_1 > 0 \quad \text{and} \quad n_2 > 0,$$
$$(\lambda_1 + \lambda_2 + \mu_1)p(n_1, 0) = \lambda_1 p(n_1 - 1, 0) + \mu_1 p(n_1 + 1, n_2)$$
$$\cdot \text{if} \quad n_1 > 0 \quad \text{and} \quad n_2 = 0,$$
$$(\lambda_1 + \lambda_2 + \mu_2)p(0, n_2) = \mu_1 p(1, n_2) + \lambda_2 p(0, n_2 - 1)$$
$$+ \mu_2 p(0, n_2 + 1)$$
$$\cdot \text{if} \quad n_1 = 0 \quad \text{and} \quad n_2 > 0,$$
$$(\lambda_1 + \lambda_2)p(0, 0) = \mu_1 p(1, 0) + \mu_2 p(0, 1).$$

We are going to describe the calculations rather than carrying them out. In fact they are long and tedious. It is necessary to calculate the generating function:

$$F(x, y) = \sum_{n_1, n_2} p(n_1, n_2)x^{n_1}y^{n_2}.$$

From the Chapman–Kolmogorov equations one finds:

$$F(x, y) = \left(\frac{1 - \rho_1 - \rho_2}{1 - \eta - y\rho_2}\right)\left(\frac{1 - \eta}{1 - \eta x}\right)$$

where

$$\eta = \frac{1}{2\mu_1}[\mu_1 + \lambda_1 + \lambda_2(1 - y) - \sqrt{[\mu_1 + \lambda_1 + \lambda_2(1 - y)]^2 - 4\lambda_1\mu_1}]$$

From the generating function we deduce:

$$E[N_1] = \left[\frac{\partial F(x, y)}{\partial x}\right]_{x = y = 1} = \frac{\rho_1}{1 - \rho_1},$$

$$E[N_2] = \left[\frac{\partial F(x, y)}{\partial x}\right]_{x = y = 1} = \frac{\rho_2 + E[N_1]\rho_2}{1 - \rho_1 - \rho_2}.$$

For the equilibrium state, calculation of the state probabilities is carried out in the following manner:

$$p(n_1, n_2) = \frac{1}{n_1!n_2!}\left[\frac{\partial^{n_1 + n_2}F(x, y)}{\partial x^{n_1}\partial y^{n_2}}\right]_{x = y = 0},$$

in particular $p(0, 0) = 1 - \rho_1 - \rho_2$. The stability condition is then given by $p(0, 0) > 0$, that is $\rho_1 + \rho_2 < 1$.

3.4.2. M/GI/1 queue with different customer classes and absolute priorities

If one is interested in the number of customers in the system or the response time only in the first moments, a simpler calculation than previously leads to the same results by extending the field of validity of this solution. We shall suppose that the queue is of type M/G/1 and that there are R classes of customer. The total arrival rate is:

$$\lambda = \lambda_1 + \lambda_2 + \cdots + \lambda_R.$$

Let μ_i^{-1} and Ks_i be the mean and the square of the variation coefficient of the service distribution of customers of class i. The mean waiting time of a customer in class i is partitioned in the following manner:

$$E[W_i] = E[W_i'] + E[W_i'']$$

where $E[W_i']$ is the mean waiting time before service and W_i'' is the waiting time before service is resumed each time it is interrupted by a customer of higher priority.

W_i': itself can be separated into three parts:
W_i^0: The residual waiting time for service of the customer if his priority is greater than or equal to that of the arriving customer.
W_i^1: The waiting time corresponding to service of customers of greater or equal priority who are present in the queue when the customer arrives there.
W_i^2: The waiting time corresponding to the service times of customers of higher priority who arrived during the waiting time in the queue of customer i.

We shall assume that service is resumed where it is interrupted.
Hence:

$$\hat{\lambda}_i = \sum_{j=1}^{i} \lambda_j, \ E[S_i] = \sum_{j=1}^{i} \frac{\lambda_j}{\lambda} \mu_j^{-1}, \quad \hat{\rho}_i = \sum_{j=1}^{i} \rho_j = E[S_i]\lambda,$$

$$E[S_i^2] = \sum_{j=1}^{i} \frac{\lambda_j}{\lambda} \frac{Ks_j + 1}{\mu_j^2};$$

these are the first two moments of the accumulated service time distribution for customers of class $1, \ldots, i$.

$\hat{\rho}_i$ is the probability that the service station is occupied and the mean waiting time to the end of this service is given by:

$$E[\bar{S}_i] = \frac{E[S_i^2]}{2E[S_i]}.$$

$1 - \hat{\rho}_i$ is the probability that the service station is not used or that it is occupied by a customer of lower priority.
We have:

$$E[W_i^0] = \hat{\rho}_i \frac{E[S_i^2]}{2E[S_i]} = \frac{\lambda}{2} E[S_i^2] = \frac{1}{2} \sum_{j=1}^{i} \frac{\lambda_j}{\mu_j^2}(Ks_j + 1).$$

The quantity $E[W_i^1]$ is the mean waiting time corresponding to the mean number of individuals that the customer arriving in the queue finds ahead of him. This mean number of customers is obtained by applying Little's formula:

$$E[N_i] = \lambda_i E[W_i'].$$

Hence:

$$E[W_i^1] = \sum_{j=1}^{i} \lambda_j E[W_j']\mu_j^{-1} = \sum_{j=1}^{i} \rho_j E[W_j'].$$

The quantity $E[W_i^2]$ is the mean service time of customers with a priority greater than i who arrive while customer i is waiting.

From Little's formula, we obtain the mean number $E[M_j]$ of customers of class j who arrive during the waiting period:

$$E[M_j] = \lambda_j E[W_i'].$$

Hence:

$$E[W_i^2] = \sum_{j=1}^{i-1} \lambda_j E[W_i']\mu_j^{-1} = E[W_i']\hat{\rho}_{i-1}.$$

Combining the three mean waiting times, we have:

$$E[W_i'] = E[W_i^0] + E[W_i^1] + E[W_i^2],$$

$$= \frac{1}{2} \sum_{j=1}^{i} \frac{\lambda_j}{\mu_j^2}(Ks_j + 1) + \sum_{j=1}^{i} \rho_j E[W_j'] + E[W_i']\hat{\rho}_{i-1}$$

From this we can deduce the value of $E[W_i']$ as a function of the mean waiting times of customers of higher priority:

$$E[W_i'] = \frac{1}{1 - \hat{\rho}_{i-1}} \left[\frac{1}{2} \sum_{j=1}^{i} \frac{\lambda_j}{\mu_j^2}(Ks_j + 1) + \sum_{j=1}^{i} \rho_j E[W_j'] \right], \quad i \geq 2.$$

This allows us to calculate the mean waiting times in successive steps knowing that:

$$E[W_1'] = \frac{1}{1 - \rho_1} \left[\frac{1}{2} \frac{\lambda_1}{\mu_1^2}(Ks_1 + 1) \right] = \frac{\rho_1}{\mu_1} \frac{1 + Ks_1}{2(1 - \rho_1)},$$

hence

$$E[W_i'] = \frac{1}{2} \frac{\sum_{j=1}^{i} \lambda_j \dfrac{(Ks_j + 1)}{\mu_j^2}}{(1 - \hat{\rho}_{i-1})(1 - \hat{\rho}_i)}.$$

The total time spent at the station by a customer of priority i is $E[W_i''] + \mu_i^{-1}$. By using Little's formula again, we obtain the mean number of customers who have been served during this period:

$$E[L_j] = \lambda_j[E[W_i''] + \mu_i^{-1}].$$

Now $E[W_i'']$ is the sum of the service times of $E[L_i]$ customers:

$$E[W_i''] = \sum_{j=1}^{i-1} E[L_j]\mu_j^{-1} = [E[W_i''] + \mu_i^{-1}]\hat{\rho}_{i-1}.$$

This allows us to obtain $E[W_i'']$:

$$E[W_i''] = \frac{\hat{\rho}_{i-1}}{1 - \hat{\rho}_{i-1}} \mu_i^{-1}.$$

Finally one obtains:

$$E[W_i] = E[W_i'] + E[W_i''],$$

$$E[W_i] = \frac{1}{(1 - \hat{\rho}_{i-1})}\left[\hat{\rho}_{i-1}\mu_i^{-1} + \frac{\sum_{j=1}^{i} \lambda_j \frac{Ks_j + 1}{\mu_j^2}}{2(1 - \hat{\rho}_i)}\right].$$

3.4.3. M/GI/1 queue with different customer classes and simple priorities

When the queue is not empty, a customer of priority i must wait for the end of the service in progress, the services of customers in classes $j = 1,\ldots,i$, who are already in the queue when he arrives and finally the services of customers in higher priority classes $j = 1,\ldots,i-1$ who arrive while he is waiting. The last two quantities have already been calculated in the preceding section. The end of the present service allows all classes of customer to intervene not only customers of class $j = 1,\ldots,i$, since priority is not absolute. The mean waiting time is of the form:

$$E[W_i] = E[W_i^0] + E[W_i^1] + E[W_i^2],$$

with:

$$E[W_i^0] = \rho_R \frac{E[S_R^2]}{2E[S_R]} = \frac{1}{2}\sum_{j=1}^{R} \lambda_j \frac{Ks_j + 1}{\mu_j^2},$$

$$E[W_i^1] = \sum_{j=1}^{i} \rho_j E[W_j], \quad E[W_i^2] = \hat{\rho}_{i-1} E[W_i].$$

Which gives:

$$E[W_i] = \frac{\sum_{j=1}^{R} \lambda_j \frac{(Ks_j + 1)}{\mu_j^2}}{2(1 - \hat{\rho}_{i-1})(1 - \hat{\rho}_i)}.$$

EXERCISES

1. Consider the M/M/1/m queue. Let $q(i)$ be the probability that a customer arriving in the queue finds i customers in the system. Show that:

$$q(i) = \frac{p(i)}{1 - p(m)}.$$

2. Consider the M/G/m/m queue. Calculate the probabilities in the stationary state.

Answer

Same results as for the M/M/m/m queue.

3. Calculate the mean number of customers in the queue for the M/M/C/m system.

Answer

$$E(L) = \frac{\rho^{C+1}p(0)}{CC!(1 - \rho/C)^2}\left[1 - \left(\frac{\rho}{C}\right)^{m-C+1} - (m - C + 1)\left(\frac{\rho}{C}\right)^{m-C}\left(1 - \frac{\rho}{C}\right)\right].$$

4.   Consider the M/M/1/m/m system which represents a set of m machines with a single repairer. Calculate the probability that one machine has broken down.

Answer

$$1 - \left[\sum_{i=0}^{m}\frac{m!}{(m-i)!}\left(\frac{\mu_1}{\mu_2}\right)^i\right]^{-1}$$

where μ_1^{-1} is the mean repair time for one machine and μ_2^{-1} is the mean time between two breakdowns.

5.   Consider the M/M/C/m/m system which represents a set of m machines with C repairers. Calculate the probability that a broken machine does not immediately find a free repairer.

Answer

$$\left[\sum_{j=C}^{m}\frac{j!}{C!C^{j-C}}\binom{m}{j}\left(\frac{\mu_1}{\mu_2}\right)^j\right]p(0),$$

μ_1 and μ_2 are defined as in the preceding exercise.

6.   Consider the $M/M^Y/1$ queue in which customers are served in groups with a size exclusively equal to K. That is, the server is interrupted if the number of customers in the queue is less than K when the preceding service finishes. Calculate the stationary probabilities of this system.

Solution

$$p(n) = p(0)\frac{1 - r_0^{n+1}}{1 - r_0}\quad\text{if } 1 \leqslant n < K,$$

$$p(n) = p(0)\frac{\lambda}{\mu}r_0^{n-K}\quad\text{if } n \geqslant K,$$

$$p(0) = \frac{1 - r_0}{K}$$

and r_0 is the smallest root of the equation

$$\mu x^{K+1} - (\lambda + \mu)x + \lambda = 0.$$

BIBLIOGRAPHY

Birth and death processes form an important class of Markov processes. In Cinlar [1] one can find very complete principles of these processes. All of Section 3.1 treated extensions to the M/M/1 queue using only birth and death processes. These developments are classic [2][3][4] and we shall make reference only to some possible extensions ([5] to [8]). The use of diffusion processes for the study of queues has been proposed by Gaver [9]. The different kinds of barrier are described by Bharudia-Reid [10]. The use of these barriers in the context of queues was proposed in [11] to [14]. The problem of discretization is considered in [14] to [17]. The use of diffusion processes to study the transient period is proposed by Kobayashi [18]. Extensions and applications are studied in [19] to [28].

Queues with priority have been much studied during the sixties and an excellent work on the subject, summarizing the principal research carried out, has been published in 1968 by Jaiswal [29]. The principal recent papers are given in references [30] to [36].

1. Cinlar, E. (1975). *Introduction to Stochastic Processes*, Prentice Hall.
2. Erlang, A. K. (1909). The theory of probabilities and telephone conversations, *Nyt Tidsskrift matematik* B. **20**, 33–39.
3. Erlang, A. K. (1917). Solution of some problems in the theory of probabilities of significance in automatic telephone exchanges, *p.o. Elec. Eng. J.*, **10**, 189–197.
4. Benes, V. E. (1957). A sufficient set of statistics for a simple telephone exchange model, – *Bell Syst. Tech. J.*, **36**, 939–964.
5. Barrer, D. Y. (1957). Queueing with impatient customers and indifferent clerks, *Oper. Res.*, **5**, 644–649.
6. Takacs, L. (1969). On Erlang's formula, *Ann. Math. Stat.*, **40**, 71–78.
7. Naor, P. (1956). On machine interference, *J. Roy. Stat. Soc. Ser.* B. **18**, 280–287.
8. Gross, D., and Harris, C. M. (1972). Continuous-review (s, S) inventory models with state dependent leadtimes. *Manag. Sci.*, **19**, 567–574.
9. Gaver, D. P. (1968). Diffusion approximation and models for certain congestion problems, *J. Appl. Prob.*, **5**, 607–623.
10. Bharudia-Reid, A. T. (1960). *Elements of the Theory of Markov Processes and their Applications*, McGraw-Hill.
11. Gelenbe, E. (1975). On approximate computer systems models, *J. ACM*, **22**, 261–269.
12. Gaver, D. P., and Shedler, G. S. (1973). Processor utilization in multiprogramming systems via diffusion approximations, *Oper. Res.*, **21**, 569–576.
13. Kobayashi, H. (1974). Application of the diffusion approximation to queueing networks, *J. ACM*, **21**, 316–328.
14. Reiser, M., and Kobayashi, H. (1974). Accuracy of the diffusion approximation for some queueing systems, *IBM Journal Res. Develop.*, **18**, 110–124.
15. Gelenbe, E. (1976). *A non-Markovian diffusion model and its application to the approximation of queueing system behaviour*, Rapport de recherche IRIA LABORIA, 158.
16. Badel, M. (1975). Quelques problèmes liés à la simulation de modèles de systèmes informatiques, Thèse de docteur-ingénieur, Université de Paris VI.
17. Yu, P. S. (1977). *On accuracy improvement and applicability conditions of diffusion approximation with application to modelling of computer systems*, – Res. Report 129, University of Stanford, USA.
18. Kobayashi, H. (1974). Application of the diffusion approximation to queueing networks: part II, – *J. ACM*, **21**, 459–469.
19. Gelenbe, E., and Pujolle, G. (1976). Approximation to single queue in a network – *Acta Informatica*, **7**, 123–136.
20. Badel, M., and Zonzon, M. (1976). *Validation d'un modèle à processus de diffusion pour un réseau de files d'attente général*, Rapport de recherche IRIA-LABORIA 209.
21. Anderson, H. A., and Sargent, R. (1972). The statistical evaluation of the performance of an experimental APL/30 system. *In Statistical Computer Performance Evaluation* W. Freiberger (ed.) pp. 73–98, Academic Press.

22. Dinh, V. (1977). *Application of a diffusion model to computer performance evaluation*, Rapport de Recherche IBM-FRANCE.
23. Gelenbe, E., and Pujolle, G. (1977). A diffusion model for multiple class queueing networks, *Proc. 3rd International Symposium Modelling Performance Evaluation Computer Systems, Bonn.*
24. Hamachmi, B., and Franta, W. R. (1978). A diffusion approximation to the multiserver queue. *Management Sciences*, **24**, 522–529.
25. Chiamsiri, S., and Moore, S. C. (1977). *Accuracy comparisons between two diffusion approximations for M/G/1 queues*, Meeting of the Oper. Res. Society and the Inst. of Manag. Sciences, Atlanta.
26. Labetoulle, J., and Pujolle, G. (1977). Modelling of packet switching communication network with finite buffer size at Each node, *Proc. IFIP WG 7, York-town Heights.*
27. Pujolle, G. (1979). The influence of protocols on the stability conditions in packet-switching network, *IEEE Trans. Com.*, **27**, 611–619.
28. Kobayashi, H., Onozato, Y., and Huynh, D. (1977). An approximate method for design and analysis of an AL OHA system, *IEEE Trans. Com*, **25**, 148–157.
29. Jaiswal, N. K. – *Priority Queues*, New York, Academic Press.
30. Jaiswal, N. K. (1961). Preemptive resume priority queue, *Operations Research*, **9**, 732–742.
31. Jackson, J. R. (1961). Queues with dynamic priority discipline, *Manag. Sciences*, **8**, 18–34.
32. Avi-Itzhak, B. (1963). Preemptive repeat priority queues as a special case of the multipurpose server problem I, *Operations Research*, **11**, 597–609.
33. Avi-Itzhak, B. (1963). Preemptive repeat priority queues as a special case of the multipurpose server problem II, *Operations Research*, **11**, 610–619.
34. Welch, P. D. (1964). On preemptive resume-priority queues, *Ann. Math. Statis.*, **35**, 600–612.
35. Chang, W. (1965). Preemptive priority queues, *Operations Research*, **13**, 820–827.
36. Greenberger, M. (1966). *The priority problem and computer time sharing, Manag. Sciences*, **12**, 888–096.

CHAPTER 4

Baskett, Chandy, Muntz and Palacios Networks

Between 1965 and 1975, Jackson networks were the only ones used for modelling and evaluating computer systems on account of their great simplicity of use. Furthermore, the exponential distribution has a fairly large application area; laws which differ from the exponential law can be replaced by the latter without making a noticeable difference to the solution.

The introduction of modelling of computer systems in the seventies restarted research into simple solutions for new networks. It is to this development that we owe the birth of Baskett, Chandy, Muntz and Palacios networks, or BCMP networks, which retain the solution in the equilibrium state in product form by introducing different classes of customer and new service disciplines.

4.1 - BCMP NETWORKS

In this paragraph we shall consider more general queueing networks than those encountered in the preceding chapter, in particular it will be possible to have a number R of classes of customer such that $R \geqslant 1$. Let N be the number of service stations. If the system is open, stations 0 and $N + 1$ are fictitious stations representing the input and output of the network respectively. Customers circulate in the network, belonging to one of the R classes. The probabilities of a step by a customer across the network are given by way of a Markov chain of transition probability:

$$P = \{p_{ir, jr'}\} \quad \text{where} \quad 0 \leqslant i \leqslant N, \quad 1 \leqslant j \leqslant N + 1, \quad 1 \leqslant r, r' \leqslant R.$$

The quantity $p_{ir, jr'}$ expresses the probability that a customer in class r at station i goes to queue j in class r'. $p_{ir, N+1r'}$, gives the probability that a customer in class r at station i leaves the network at the end of his service.

We shall state the BCMP theorem in Section 4.1.3. Before then, let us introduce the service time distributions and the service disciplines which will be used.

103

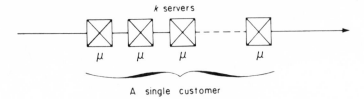

k servers

μ μ μ μ

A single customer

Figure 4.1 Erlang-k distribution

4.1.1. Service time distributions

In the Jackson model exponential probability distributions are the only ones used and they possess the following property: the time already spent at a server which obeys an exponential law gives no indication of the service still to be carried out. This model is Markovian. To conserve this last property, several extensions are possible. We shall describe them in the following paragraphs. The Erlang-k model is formed from a series of k identical exponential servers with service time μ^{-1}. A single customer moves through this model (Figure 4.1). The probability density is of the form:

$$f(t) = \frac{\mu^k (t)^{k-1} e^{-\mu t}}{(k-1)!}$$

and its Laplace transform:

$$\mathcal{L}(f(t)) = \left(\frac{\mu}{\mu + s}\right)^k.$$

The variation coefficient of this distribution has the value $1/k$ and is always less than 1.

One can imagine a generalization of the Erlang-k law in which each service point has a different rate $\mu_i, i = 1, \ldots, k$. But from its Laplace transform:

$$\prod_{i=1}^{k} \frac{\mu_i}{\mu_i + s},$$

one finds a variation coefficient which is always less than 1.

In order to obtain a variation coefficient greater than 1, the service points are arranged in parallel.

The hyperexponential model is formed from k servers in parallel as represented in Figure 4.2. A customer arriving at the station will be served at point i with a probability α_i.

The probability density is a combination of negative exponentials of the form:

$$f(t) = \sum_{i=1}^{k} \alpha_i \mu_i e^{-\mu_i t}.$$

From its Laplace transform:

$$\mathcal{L}(f(t)) = \sum_{i=1}^{k} \alpha_i \frac{\mu_i}{\mu_i + s},$$

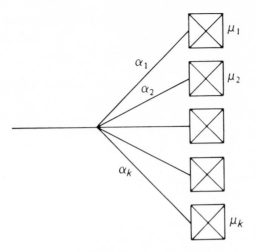

Figure 4.2 Hyperexponential model

one obtains the variation coefficient:

$$\frac{2\sum\limits_{i=1}^{k}\dfrac{\alpha_i}{\mu_i^2}}{\left(\sum\limits_{i=1}^{k}\dfrac{\alpha_i}{\mu_i}\right)^2}-1$$

which verifies the Cauchy–Schwartz inequality

$$\frac{2\sum\limits_{i=1}^{k}\dfrac{\alpha_i}{\mu_i^2}}{\sum\limits_{i=1}^{k}\alpha_i\dfrac{\alpha_i}{\mu_i^2}}-1 > 1.$$

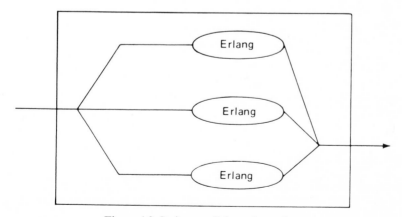

Figure 4.3 Series-parallel service points

Hence, unlike the Erlang model, the variation coefficient is always greater than 1.

One possibility of obtaining any variation coefficient is to arrange Erlang servers in parallel as in Figure 4.3.

Another possibility for obtaining more general service laws has been proposed by Cox [5]. The service point consists of a set of exponential servers, arranged as shown in Figure 4.4. A new customer can enter into service only when the preceding customer has left having cleared a certain number of stages.

To introduce the notation which we shall use in the BCMP theorem at this point, we assume that we study the ith service point and that it can be used by different classes of customer $r = 1, \ldots, R$.

We designate the total number of servers in Cox's distribution by ℓ_{ir} for a customer in class r at station i. The service time (exponentially distributed) of the mth stage of Cox's model will be denoted by μ_{irm} for a customer in class r at station i. Finally, A_{irm} will be the probability that a customer in class r at station i reaches the mth stage. In Figure 4.4, $A_{irm} = b_0 b_1 b_2 \cdots b_{m-1}$.

The Laplace transform of Cox's distribution is:

$$\mathscr{L}[f(t)] = b_0 + \sum_{m=1}^{\ell_{ir}} A_{irm}(1 - b_m) \prod_{j=1}^{m} \frac{\mu_j}{\mu_j + s}.$$

If the customer passes directly from the input to the output with probability $1 - b_0$, the service density is a Dirac function. Cox has shown that adding the possibility of skipping certain service points or, in contrast, returning to previous ones, does not increase the generality.

Comment

1. Cox's model has the Markov property (in a multidimensional state space).
2. Cox distributions are identical to distributions which have a rational Laplace transform. It should be noted that densities exist whose Laplace transforms are rational with complex poles as in the following example:

$$\mathscr{L}[f(s)] = \frac{\mu(\mu^2 + \omega^2)}{(s + \mu)((s + \mu)^2 + \omega^2)}, \quad \omega \in \mathbb{R}$$

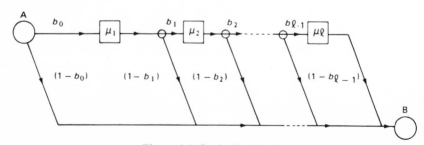

Figure 4.4 Cox's distribution

of which the density is

$$f(t) = \frac{\mu}{\omega^2} (\mu^2 + \omega^2) e^{-\mu t} (1 - \cos \omega t).$$

These complex poles correspond to fictitious service points which add no generality to Cox distributions with real poles.

3. Erlang distributions, hyperexponential and series-parallel, are Cox distributions.

4. A sum of Cox distributions is a Cox distribution.

An important property of Cox distributions is to be close to the set of distributions, that is any distribution can be approached as closely as one wishes by a Cox distribution. In fact, one can find a staircase function which approaches as closely as required to the function $f(t)$ to be replaced:

$$f(t) = \sum_{i=1}^{m} \alpha_i 1_{[a_i, b_i]}$$

where a_i and b_i are the steps in the staircase function and α_i is the interval between steps. The function $\alpha_i 1_{[a_i, b_i]}$ is an Erlang-∞ function which can be approached as closely as one wishes by an Erlang-n function with n sufficiently large. As a sum of Erlang functions is a Cox function, our function $f(t)$ can be approached as closely as one wishes by a Cox distribution.

4.1.2. Service disciplines

Service stations in a BCMP (Baskett, Chandy, Muntz, Palacios) network can obey any of the four following possibilities:

Type 1 The service discipline is first in, first out (FIFO), the station has a single server and the service time is exponentially distributed with the same mean for all classes of customer. If station i has such a server, we denote the rate of service by $\mu_i(k_i)$ if there are k_i customers in the station (including the one which is being served).

Type 2 The discipline is that of time division ('processor sharing', PS), that is, a customer at the station receives $1/k$ seconds of service per second if k is the number of customers at the station. All customers receive a small portion of their respective service time in turn. These quanta of service received on each visit to the server tend to 0. The service time distribution can be a distinct Cox distribution for each class of customer.

Type 3 The number of servers at the station is sufficient for there to be always at least one free. This leads to the fact that a new customer entering the station starts his service immediately. The service time distributions can be distinct Cox distributions for each class of customer.

Type 4 The service discipline is 'last arrived, first served' with an absolute priority for the newly arriving customer. There is a single server, that is a new arrival at the station interrupts the customer's service in order to start his own. The displaced customer is returned to the head of the queue and he rejoins his service where he left it, when the customer who caused the interruption finishes his service. As in the two preceding types of station, the probability distribution of the service time can be Cox and it can be distinct for each class of customer.

4.1.3. BCMP theorem

Let $i = 1, 2, \ldots, I$, be the indices of a partition of the classes of customer, 1, 2, ..., R; for example we could have: 1 for classes 1 and 6, 2 for class 3, 3 for classes 2, 4, 5, 7, ..., R. Again let K_i be the number of customers in the element of the partition having index i.

The arrival process of new customers in the network is Poisson with the two following possibilities:

* $\lambda(K)$ is the rate of arrival from the exterior when there are K customers in the network.

* $\lambda_i(K_i)$ is the rate of arrival from the exterior of customers in partition i of which there are K_i in the network.

Two classes of customer must belong to the same partition if customers of one class can pass into another class. The arrival rate of these partitions can depend on the number of customers K_i in each partition. If one class of customer can neither enter nor leave it forms its own section of the partition: it is a closed sub-chain.

Finally let us designate the solution for the system by e_{ir}:

$$e_{ir} = \sum_{r'=1}^{R} \left(\sum_{j=1}^{N} e_{jr'} p_{jr', ir} + p_{0r', ir} \right),$$

for $1 \leq i \leq N$, and $1 \leq r \leq R$. The quantity e_{ir} gives the relative frequency of the number of visits to station i by a customer of class r.

$\hat{k} = (\hat{k}_1, \ldots, \hat{k}_N)$ is the state vector of the network which depends on the type of service station:

Type 1

$$\hat{k}_i = (k_{i1}, \ldots, k_{ik_i})$$

where k_{ij} is the class of the jth customer waiting at station i in first come, first served order.

Type 2 or 3

$$\hat{k}_i = ((k_{i1}, s_{i1}), \ldots, (k_{ik_i}, s_{ik_i}))$$

where k_{ij} is the class of the jth customer waiting in the order of arrival and s_{ij} is the stage of the Cox model.

Type 4

$$\hat{k}_i = ((k_{i1}, s_{i1}), \ldots, (k_{ik_i}, s_{ik_i}))$$

where k_{ij} and s_{ij} are identical to those defined for types 2 and 3, the order of the k_{ij} being defined by the last arrived, first served service discipline.

Before stating the BCMP theorem, we shall write the notation used:

$$\text{Let } f_i(\hat{k}_i) = \begin{cases} \displaystyle\prod_{j=1}^{k_i} e_{ik_{ij}}/\mu_i(j) & \text{if } i \text{ is of type 1,} \\[2ex] \displaystyle k_i! \prod_{r=1}^{R} \prod_{m=1}^{r_r} \frac{(e_{ir}A_{irm}\mu_{irm}^{-1})^{k_{irm}}}{k_{irm}!} & \text{if } i \text{ is of type 2,} \\[2ex] \displaystyle\prod_{r=1}^{R} \prod_{m=1}^{r_{ir}} \frac{(e_{ir}A_{irm}\mu_{irm}^{-1})^{k_{irm}}}{k_{irm}!} & \text{if } i \text{ is of type 3,} \\[2ex] \displaystyle\prod_{j=1}^{k_i} e_{ik_{ij}}A_{ik_{ij}}\mu_{ik_{ij}}^{-1}s_{ij} & \text{if } i \text{ is of type 4,} \end{cases}$$

and

$$d(K) = \begin{cases} \displaystyle\prod_{m=0}^{K-1} \lambda(m) & \text{if the network is open,} \\[2ex] 1 & \text{if the network is closed.} \end{cases}$$

In the case of distinct arrival processes for the sub-chains $1, 2, \ldots, n$, we have:

$$d(K) = \begin{cases} \displaystyle\prod_{j=1}^{n} \prod_{i=0}^{K_n-1} \lambda_i(m) & \text{if the network is open,} \\[2ex] 1 & \text{if the network is closed.} \end{cases}$$

Let:

$$\gamma = \sum_{\hat{k}} d(K) \prod_{i=1}^{N} f_i(\hat{k}_i).$$

BCMP theorem

If $\gamma < +\infty$, an equilibrium state exists determined by

$$p(\hat{k}) = \gamma^{-1} d(K) \prod_{1}^{N} f_i(\hat{k}_i).$$

A number of comments arise:

1. The expression of probabilities in the stationary state involves only the means of the service time distributions, even if the distribution is a Cox one. Only knowledge of the means is necessary in BCMP networks.

2. The joint probability is the product of the marginal probabilities, which allows the network to be studied station by station and also facilitates the development of effective algorithms for calculation.

3. The principal difficulty encountered is calculation of the normalization constant γ. In fact, if the network has many stations having service time distributions with numerous stages, calculation of the normalization constant becomes tedious.

4. The essential advantage in comparison with the Jackson network is the existence of different classes of customer and of service disciplines other than FIFO.

5. Capacity limitations, either global or by class of customer, can be introduced. They change only the calculation of the normalization constant.

We now give some simple versions of this theorem by assuming that the service rates do not depend on the number of customers present in their respective queues.

The equilibrium state is uniquely characterized by the vector

$$\overline{k} = (\overline{k}_1, \overline{k}_2, \ldots, \overline{k}_n) \quad \text{with} \quad \overline{k}_i = (\overline{k}_{i2}, \ldots, \overline{k}_{iR})$$

where k_{ir} is the number of customers of class r in queue i. Hence we have:

$$p(\overline{k}) = \gamma^{-1} \, d(K) \prod_{i=1}^{n} g_i(\overline{k}_i)$$

where γ and $d(K)$ are identical to the values already given above and:

$$g_i(\overline{k}_i) = \begin{cases} k_i! \displaystyle\prod_{r=1}^{R} \frac{1}{k_{ir}!} \left(\frac{e_{ir}}{\mu_i}\right)^{k_{ir}} & \text{if } i \text{ is of type 1,} \\[2ex] k_i! \displaystyle\prod_{r=1}^{R} \frac{1}{k_{ir}!} \left(\frac{e_{ir}}{\mu_{ir}}\right)^{k_{ir}} & \text{if } i \text{ is of type 2 or 4,} \\[2ex] \displaystyle\prod_{r=1}^{R} \frac{1}{k_{ir}!} \left(\frac{e_{ir}}{\mu_{ir}}\right)^{k_{ir}} & \text{if } i \text{ is of type 3.} \end{cases}$$

This formulation can be further simplified by assuming that the network is open and the arrival rates from the exterior are independent of the number of customers in the system. The system state is determined by the vector $k = (k_1, k_2, \ldots, k_n)$ where k_i is the number of customers at station i.

We have

$$p(k) = \prod_{i=1}^{N} h_i(k_i),$$

where

$$h_i(k_i) = \begin{cases} (1 - \rho_i)\rho_i^{k_i} & \text{if } i \text{ is of type 1, 2 or 4} \\[2ex] e^{-\rho_i} \dfrac{\rho_i^{k_i}}{k_i!} & \text{if } i \text{ is of type 3,} \end{cases}$$

where ρ_i is the utilization rate of the server of queue i, given by

$$\rho_i = \sum_{r=1}^{R} \frac{\lambda e_{ir}}{\mu_{ir}}.$$

If one is particularly interested in networks using the FIFO service discipline, it is necessary to recognize that the contribution of BCMP networks by comparison with Jackson networks is very little: the distribution of service times is always exponential; if several classes of customer are possible, they must all have the same service time at a given station. For this particular case, which is possibly the most important, we shall introduce a more general method, giving only approximate results, in the next chapter.

To establish the proof of the BCMP theorem, it is necessary to verify that the solution satisfies the series of equations for the global balance of the system in the equilibrium state, which can be written in the following manner:

$p(\hat{k})$ [probability of leaving state $\hat{k}$]

$= \sum_{k'} p(\hat{k})$ [probability of going from state $\hat{k}'$ to $\hat{k}$].

Unfortunately these equations are particularly complicated since they also include the different classes of customer and the stages of Cox laws. Decomposition of the global balance equations into a series of equations called local balance equations has led to the solution. They are determined by:

$p(\hat{k})$ [probability that a customer of class r leaves stage ℓ of service station i]

$= \sum_{k'} p(k')$ [probability that a customer of class r enters stage ℓ of

service station i]

and

$p(\hat{k})$ [probability that a customer of class r leaves station i]

$= \sum_{k'} p(\hat{k})$ [probability that a customer of class r enters station i]

If one sums the series of local balance equations over the classes of customer and the stages of the Cox distribution, one obtains the global balance equations again. If the local balance equations allow a solution to be obtained, it is evident that this solution will also be that of the global balance equations, but the converse is false.

A specific example of decomposition into local balance equations is given in Chapter 6. One can easily show that the solution expressed in the BCMP theorem is effectively verified by the local balance equations.

This theorem which arose from studies made relatively independently by the four authors has been slightly extended since its first formulation (1975). The essential step has been to attempt to understand the basic reason for the product form. Firstly, stations have the property of conserving the Markovian character of the interarrivals; if a Poisson process is applied at the input of one of the four types of station described in the BCMP theorem, the output process is also Poisson and independent of the input process. This characteristic is called the $M => M$ property introduced by Muntz in 1972.

It cannot be concluded from this remark that streams in a BCMP network are all Poisson processes. In Chapter 6 we shall show that most streams are neither Poisson nor renewal processes.

It can be shown that a Markovian network which satisfies the local balance equations has the $M => M$ property. Furthermore, Chandy, Howard and Towsley show that a Markovian network, which is in product form, has all its local balance equations satisfied. If there are several classes of customer, the disciplines among the classes must be independent.

Another property of BCMP type networks has been observed by Sevcik and Mitrani [21] and also by Lavenberg and Reiser [22]; when a customer leaves a queue, he sees the network in its equilibrium state in the case of an open network and in its equilibrium state losing a customer (himself) in the case of a closed network.

A generalization of the BCMP theorem has been introduced by Kelly [7], allowing customers to follow arbitrary paths in the network and not to be guided by Bernouilli branches. Also Kelly [9] showed with Barbour [8] that where Cox distributions are found it is possible to put general service distributions.

In studying punctual processes formed from input streams at the stations, the East German School has shown a property of 'insensitivity' for a certain type of queue; the solution at the stationary state of a station depends only on the first moment of the service time probability distribution. Further to the four types of BCMP station which evidently satisfy this property, there is also the $./GI/m/m$ queue where customers lost due to the limit of the queue are served for zero time and continue in the network.

Other extensions are allowed by limiting the number of customers in the global network or in certain sub-chains of the network. Lam [11] shows that the product form is conserved if a customer leaving the network creates an entry into the network. Pellaumail [23] and Le Ny [24] introduce networks in which the probabilities of progress depend on the state of the network; the product form is conserved conditional on an assumption of 'class exchangeability'. Finally, to conclude the extensions, a very important property which allows the product form to be explained in another way is given by conservation of the stream distributions; the probability distribution of the intervals between arrivals is equal to the probability distribution of the intervals between departures for each station of the network. Following a previous remark, it should be noted that these are generally not Poisson. A highly schematic explanation of the product form depends on the fact that suppression of one station of the network does not modify the behaviour of the streams; since the stations are independent, the joint probability must be in product form.

4.1.4. Examples of BCMP networks

Consider a network in which the path of a customer depends on his past, that is a customer at station i_n will be directed to a destination depending on the stations already visited i_{n-1}, i_{n-2}, $\ldots$, i_1.

To study such a network, it is sufficient to introduce additional classes of customer to take account of the possible paths. In particular, if the path is determined in advance, it is necessary to define a class of customer for each series of stations from entry to departure. Notice that customers of various classes must have the same service distribution for the BCMP theorem to be applicable if the service discipline is first in, first out. This will be verified in the example which we develop later.

4.2. - RESPONSE TIME OF A PACKET SWITCHING NETWORK

A more complete description of a packet switching network is presented in Section 5.6. Here we only need to know that a network of this kind directs customers from an input point to an exit point by passing through a certain number of stations which are predefined by the origin and destination of the

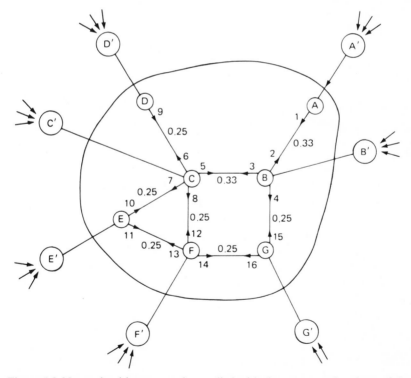

Figure 4.5 Network with seven nodes studied with the mean service times of the queues numbered 1 to 16

customer. These customers (or packets which carry the data) circulate in a network consisting of seven nodes following the topology described in Figure 4.5. The route followed by the customers is always the shortest. In the case where several possibilities exist, the choice will be made in accordance with the path given in Table 4.1. For example, customers entering at A′ and heading towards F′ will pass through nodes A, B, C, F and traverse queues 1, 3 and 8. The queue from F to F′ is negligible; in fact output lines from a node to the exterior are much faster than internal lines. The pathways and the topology define 42 classes of customer. The packets are assumed to have an exponentially distributed length, which implies that the service times are also exponentially distributed since the speed of the lines is constant. The mean service time is indicated in Figure 4.5; 0.25 second or 0.33 second following the lines.

The BCMP theorem allows us to analyse the behaviour of the network from performance criteria. We can also calculate the global response time, the utilization of servers and the mean number of customers in the network. By

Table 4.1

Source	Destination	Via
A′	F′	C
B′	F′	C
C′	G′	F
D′	G′	F
G′	C′, D′	B
F′	A′, B′	C

Table 4.2 - **Mean number of customers in the 16 queues of the network represented in Figure 4.5 for two values of λ**

Queue	$\lambda = 0.143 = 1/7$	$\lambda = 0.214 = 1.5/7$
1	0.428	0.828
2	0.428	0.818
3	1	3
4	0.154	0.250
5	0.667	1.5
6	0.667	1.5
7	0.464	0.66
8	0.667	1.5
9	0.667	1.5
10	0.364	0.66
11	0.154	0.25
12	0.364	0.66
13	0.154	0.25
14	0.364	0.66
15	0.364	0.66
16	0.154	0.25

varying parameters such as input rate to the network, the path or the service times, one obtains an immediate response by using a computer program. By way of example, we have calculated the mean number of customers per station by assuming that the arrival rate is identical for each class of customer and equal to 0.143 at first and then 0.214. The results are given in Table 4.2.

BIBLIOGRAPHY

The first signs permitting the assumption that Jackson networks could be extended appeared in 1972 in the papers of Chandy [2] and Muntz [3]. They introduced the local balance equations and the $M = > M$ property. Baskett and Palacios [4] were interested in the case of a partitioned processor during the same year. Cox distributions [5] started to be used with new service stations. The BCMP theorem [1] appeared in 1975 in its definitive version. Reiser and Kobayashi [6] and also Kelly [7] give slightly different versions, but Kelly introduces it in a quite different manner which allows a certain number of extensions [8] [9]. From the approach of Baskett, Chandy, Muntz and Palacios, extensions are proposed by Gelenbe and Muntz [10], Lam [11] then Chandy, Howard and Towsley [12].

The East German School brought new extensions by using 'insensitivity' and punctual processes; Schassberger [13] [14] and Stoyan [15]. The explanation of the product form is researched from the properties of streams: Beutler and Melamed [16], Melamed [17], Labetoulle, Pujolle and Soula [18], Pujolle [19], Brémaud [20]. The property that the BCMP network is in equilibrium when a customer leaves is shown by Sevcik and Mitrani [21] and Lavenberg and Reiser [22]. Finally, the product form can be conserved by introducing path probabilities which depend on the state, Pellaumail [23] and Le Ny [24].

1. Baskett, F., Chandy, K. M., Muntz, R. R., and Palacios, F. G. (1975). Open, closed and mixed networks of queues with different classes of customers, *J. ACM*, **22**, 248–260.
2. Chandy, K. M. (1972). The analysis and solutions for general queueing networks, *Proc. Sixth Annual Princeton Conference Information Sciences Systems, Princeton University*.
3. Muntz, R. R. (1972). Poisson departure processes and queueing Networks, IBM Research Report, RC 4145, IBM Thomas J. Watson Research Center, Yorktown Heights, New York.
4. Baskett, F. and Palacios, F. G. (1972). Processor sharing in a central server queueing model of multiprogramming with applications, *Proc. 6th Ann. Princeton Conf. Information Science Systems*, pp. 598–603. Princeton, New Jersey.
5. Cox, D. R. (1955). A use of complex probabilities in the theory of stochastic processes, *Proc. Cambridge Phil. Soc.*, **51**, 313–319.
6. Reiser, M. and Kobayashi, H. (1975). Queueing networks with multiple closed chains: theory and computational algorithms, *IBM J. Res. and Dev.*, **19**, 283–294.
7. Kelly, F. P. (1975). Networks of queues with customers of different types, *J. Appl. Prob.*, **12**. 542–554.
8. Barbour, A. D. (1976). Networks of queues and the method of stages, *Adv. Appl. Prob.*, **8**, 584–591.
9. Kelly F. P. (1976). Networks of queues, *Adv. Appl. Prob.*, **8**, 416–432.
10. Gelenbe, E. and Muntz, R. R. (1976). Probabilistic models of computer systems, *Acta Informatica*, **7**, 35–60.
11. Lam, S. S. (1976). Store- and-forward buffer requirements in a packet switching network, *IEEE Trans. Com.* **24**, 394–399.

12. Chandy, K. M., Howard, J. K., and Towsley, D. F. (1977). Product form and local balance in queueing networks, *J. ACM* **24**, 250–263.
13. Schassberger, R. (1978). insensitivity of steady-state distributions of generalized semi-markov processes with speeds, *Adv. Appl. Prob.*, **10**, 836–851.
14. Schassberger, R. (1978). The insensitivity of stationary probabilities in networks of queues, *Adv. Appl. Prob.* **10**, 906–912.
15. Stoyan, D. (1978). Queueing networks, insensitivity and a heuristic approximation, elektion, *Informations verarbeit, Kybernetik*, **14**.
16. Beutler, F. J. and Melamed, B. (1978). Decomposition and customer streams of feedback networks of queues in equilibrium, *Operations Research*, **26**, 1059–1072.
17. Melamed, B. (1979). On poisson traffic processes in discrete-state Markovian systems with applications to queueing theory, *Adv. Appl. prob.*, **11**, 218–239.
18. Labetoulle, J., Pujolle, G., and Soula, C. (1981). Distribution of the flows in a general Jackson network, *Math. Oper. Res.*, **6**, 173–185.
19. Pujolle, G. (1980). Réseaux de files d'attente à forme produit, *RAIRO, Recherche Opérationnelle* **14**, 317–330.
20. Bremaud, P. (1981). Dynamical Point Processes and Ito Systems in Communications and Queueing, Springer-Verlag.
21. Sevcik, K. and Mitrani, I. (1979). The distribution of queueing network states at input and output instants, *Proc. Int. Symp. performance Computer Systems*, Vienne.
22. Lavenberg, S. and Reiser, M. (1979). The state seen by an arriving customer in closed multiple chain queueing networks, Rap. Rech. IMB. Yorktown Heights.
23. Pellaumail, J. (1978). Régimes stationnaires quand les routages dépendent de l'état, *Actes du 1er colloque AFCET-SMF de mathématiques appliquées*, Palaiseau.
24. Le Ny, L. M. (1979). Étude analytique de réseaux de files d'attente multi-classes à routages variables. *Thèse du 3ème cycle, Rennes I*.

CHAPTER 5

Approximate Methods

5.1 - INTRODUCTION

The only queueing networks for which the solution is actually known in an explicit form are those described in the preceding chapters—Jackson networks and Baskett–Chandy–Muntz–Palacios networks. If one wishes to study very large networks with several hundreds of queues, calculation of the normalization constant can pose problems and can even become impossible. It may be noticed that if one is interested only in the first come, first served discipline, networks of more than three queues for which the solution is known have the following characteristics:

–the service time distributions are negative exponentials,
–the external arrivals are Poisson,
–there is only one class of customer or, if there are several, the service times are independent of the class,
–all queues have unlimited capacity.

It can be seen immediately that the number of cases which can be resolved is extremely small. Furthermore, solutions known explicitly give joint probabilities which are not necessarily what we need in practice. Given the difficulty and noticing the impossibility of obtaining exact solutions in numerous cases, in this chapter we examine a number of approximate methods which allow a solution to be approached.

To introduce this chapter, an attempted classification of the different methods used for mathematical modelling of systems of queues is given in Figure 5.1. The procedure involves the following phases:

–analysis of the real system which leads to a *model* (a queueing network);
–formulation of a set of *equations* which govern the model. The parameters which characterize the model (service distributions, input stream distributions, ...) are obtained by measurement or derived from knowledge of the real system;
–solution of the set of equations.

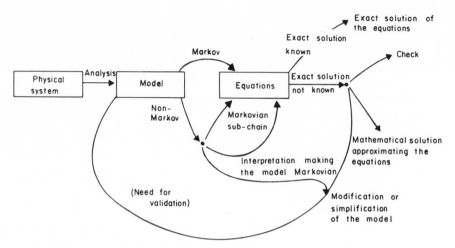

Figure 5.1 Classification of the analysis methods for queueing systems

The analysis phase which allows a queueing model to be obtained from a physical system is not based on a general method. Above all it is experience which, at this stage, enables the definition of 'good models' in respect of their precision, their level of detail and the existing methods of solution. The second phase, transition from a queueing model to equations, that is a properly specified mathematical model, is much more difficult. It is first necessary to define the set of system states from the model, then to calculate the relationships which exist between the probabilities of these states. This is possible in general only if the model is Markovian; the changes of state depend only on the present state and not on the history of the system. If this is not the case, one resorts to one of the four following approaches:

–to define embedded sub-chains which are, themselves, Markovian. We have seen that M/G/1 and G/M/1 queues are studied in this way;
–to make the model Markovian by creating additional, virtual, states. We have seen an example in decomposition by Cox's distribution. Another example is decomposition of an Erlang-n distribution into n exponential services in tandem;
–to simplify the model, for example by assuming that a server is exponential, that a stream is Poisson or that a queue is unlimited. In this case, as in the following, it is desirable to estimate the maximum error if possible or at least to have an idea of the deviation from the exact solution;
–to modify the model so that it can be used by omitting or adding certain elements.

In this chapter, we are going to consider several approximate methods: firstly, a decomposition method which consists of studying each queue of the network in turn with characterization of intermediate streams, followed by a mean value analysis which has the sole purpose of studying the first moments of the mean numbers in each queue. Then we shall see the aggregation method which is a study of the system group by group; these are determined by weak interactions with the exterior. Finally, we shall examine the isolation method which consists of isolating each queue and studying it formally taking account of the effects of interaction with the exterior. An explicit solution is then obtained for the whole system.

Several other approximate methods exist but they generally lead to numerical solutions which require long calculation times on a computer. One can quote numerical methods using the transition rate matrix. The stationary probability is obtained from the eigen vectors of the matrix. The principal difficulty or constraint is the number of system states. Only very small systems can be studied. Other interesting approximate methods are represented by iterative methods. In general, for each station, it is necessary to find a set of arrival rates such that the behaviour of the station becomes closer and closer to the actual behaviour. One decides that the calculated behaviour is close to the (unknown) actual behaviour when certain conditions are satisfied (problems of stop tests). The difficulty with these iterative methods is the uncertainty of stop tests corresponding to the exact solution. Also the algorithms are only rarely convergent.

5.2 - DECOMPOSITION METHOD

Each station of an open network contains a single server whose service discipline is exclusively first in, first out. Knowledge of the joint probability is not in general very useful. Only the mean number of customers at each station of the network and the response time are of interest to users. The method which we are going to develop allows measures of these two parameters to be obtained. It consists of studying each queue of the network by a queue by queue

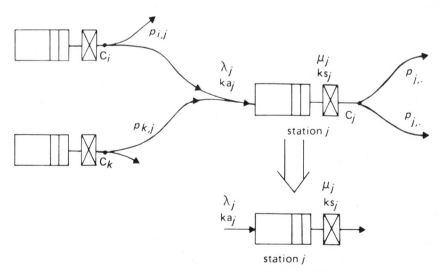

Figure 5.2 Decomposition method consisting of calculating the values of λ_j and ka_j

decomposition of the network. The distributions of service times and time intervals between successive arrivals are characterized by their first moments or more exactly by their rates and the square of their variation coefficient (the variance multiplied by the rate squared). We have attempted to represent the method in Figure 5.2.

This is based on the following generally unjustified assumption: the departure process from any station is a renewal process: the time interval between two departures does not depend on the preceding intervals. This assumption is exact in the case of Poisson arrivals and exponentially distributed services or when the station is saturated.

Let us call C_i the square of the variation coefficient (SVC) of the intervals between two successive departures from station i. Let us again denote A_i, S_i and τ_i the random variables representing the time interval between two arrivals, the service time and the time interval between two successive departures respectively all for station i.

A good approximation to the time interval between two successive departures is given by:

$$\tau_j = \begin{cases} S_j & \text{with the probability } \rho_j = \lambda_j/\mu_j, \\ A_j + S_j & \text{with the probability } 1 - \rho_j. \end{cases}$$

From which we deduce:

$$E[\tau_j] = \mu_j^{-1}\rho_j + (1 - \rho_j)(\lambda_j^{-1} + \mu_j^{-1}) = \lambda_j^{-1}$$

as expected, and

$$E[\tau_j^2] = \lambda_j^{-2}(1 + C_j) = E[S_j^2] + (1 - \rho_j)(E[A_j^2] + 2E[A_j]E[S_j]),$$

which implies:

$$C_j = -1 + \rho_j^2(Ks_j + 1) + (1 - \rho_j)(\lambda_j^2 E[A_j^2] + 2\rho_j).$$

We obtain a relation between C_j and $E[A_j^2]$. To determine $E[A_j^2]$ we are going to calculate the variance, Var_j, of the number of customers entering station j during a time t, by assuming that the processes of each queue in the network are independent. The variance being the sum of the variances, we obtain:

$$\text{Var}_j = \sum_{i=0}^{n} [(C_i - 1)p_{ij} + 1]\lambda_i p_{ij}t;$$

thanks to a limit theorem for renewal processes we also have:

$$\text{Var}_j = \lambda_j^3 [E[A_j^2] - \lambda_j^{-2}]t.$$

These last two equalities determine the value of $E[A_j^2]$ which when substituted in the relation between C_j and $E[A_j^2]$ gives:

$$C_j = -1 + \rho_j^2(Ks_j + 1) + (1 - \rho_j)(2\rho_j + 1 + \frac{1}{\lambda_j} \sum_{i=0}^{n} [(C_i - 1)p_{ij} + 1]\lambda_i p_{ij}).$$

The solutions of this system of linear equations give the values of $C_j, j = 1, \ldots, n$.

The square of the variation coefficient of interarrivals at station j can be written in the form:

$$Ka_j = \frac{1}{\lambda_j} \sum_{i=0}^{n} [(C_i - 1)p_{ij} + 1]\lambda_i p_{ij}.$$

Knowledge of the values of C_i, $i = 0, 1, \ldots, n$, allows the values of Ka_i, $i = 1, \ldots, n$ to be obtained.

Comment

C_j can also be written in the form:

$$C_j = -1 + \rho_j^2(Ks_j + 1) + (1 - \rho_j)(2\rho_j + 1 + Ka_j)$$

which is a relation giving the value of the variation coefficient of the time intervals between departures from a station when one knows the arrival and service processes by their first two moments.

Station j can now be studied using the values of the first two moments of the arrival and service processes.

In particular, the mean length of the queue is well approximated by Kingman's formula (Section 1.5) and the response time is obtained by application of Little's formula (Section 1.4).

In summary, the solution of two linear systems, one giving the values of visiting rates at various stations e_j, and the other giving the variation coefficient of the intervals between two successive departures C_j, are sufficient to obtain a solution for a very general queueing network. We are going to give several applications of this method (cf. Section 5.6). As before we are going to extend it by assuming that there are several different classes of customer.

Assume that each station has a single server who applies the first in, first out discipline without priority. The rth class of customer, $1 \leqslant r \leqslant R$, is characterized by:

$-$a stream of customers arriving from the exterior with arrival rate λ_{0r} and a variation coefficient whose square is Ka_{0r}. We shall put:

$$\lambda = \sum_{r=1}^{R} \lambda_{0r};$$

$-$service times of rate $\mu_{j,r}$ and a square of variation coefficient $Ks_{j,r}$ at the jth station.

As in Section 4.2, the probabilities of going from one station to another are determined by a Markov chain of transition probability $p = \{p_{ir, jr'}\}$. The relative frequency of the number of visits by a customer of class r to station i is given by e_{ir}, which is the solution of the system:

$$e_{ir} = \sum_{r'=1}^{R} \left(\sum_{j=1}^{N} e_{jr'} p_{jr', ir} + p_{0r', ir} \right).$$

Let $\rho_{i,r} = \lambda e_{ir}/\mu_{ir}$ be the value of the load at station j due to customers of class r. Let us again define:

$$\rho_j = \sum_{r=1}^{R} \rho_{jr}, \quad \lambda_j = \sum_{r=1}^{R} \lambda_{jr} \quad \text{and} \quad \Pi_{jr} = \frac{\lambda_{jr}}{\lambda_j}.$$

The quantity ρ_j is the utilization rate of station j and λ_j is the entry rate at station j for all classes. The arrivals and departures of customers of class r, $1 \leqslant r \leqslant R$, are assumed to form a renewal process. Using a similar argument to that of the preceding paragraph, we obtain a system which determines the values of the C_j and the squares of the variation coefficients of the time intervals between departures from station j, $1 \leqslant j \leqslant K$:

$$C_j = -1 + \lambda_j \sum_{r=1}^{R} \rho_{jr}\mu_{jr}^{-1}(Ks_{jr} + 1) + (1 - \rho_j)(Ka_j + 1 + 2\rho_j)$$

where

$$Ka_j = \lambda_j^{-1} \sum_{k=0}^{m} [(C_k - 1)P_{k_j} + 1]\lambda_k p_{kj},$$

with

$$P_{kj} = \sum_{r=1}^{R} \sum_{r'=1}^{R} \frac{\lambda_{kr}}{\lambda_k} p_{kr,jr'}.$$

The square of the variation coefficient of the interarrivals at station j for customers of class r is hence:

$$Ka_{jr} = (Ka_j - 1)\Pi_{jr} + 1.$$

Knowing λ_j and Ka_j, station j can be characterized by the mean length of its queue, given by Kingman's formula:

$$E[N] = \rho_j \left[1 + \frac{\rho_j(Ka_j + Ks_j)}{2(1 - \rho_j)} \right].$$

The response time is obtained by applying Little's formula $E[T] = E[N]/\lambda$.

5.3 - MEAN VALUE METHOD

Mean values such as those of response times and the number of customers in a queue, are always the characteristics most required of a model. So the method which we are now going to present consists of calculating these means.

The decomposition method which we described in Section 5.2 applies to open networks. We now describe a method which applies to two types of network, open and closed.

We shall assume firstly that the service discipline in each queue of the network is first in, first out (FIFO). There are always n stations in the network and we shall assume that there is only one class of customer. This last assumption will be abandoned later.

The basic idea is very simple to explain; the mean response time of a station

i of a general network is simply the sum of the mean service times and the mean waiting times before being served. Let $E[N_i^*]$, $E[T_i]$ and $E[S_i]$ be the mean number of customers in the queue at the instant of arrival, the mean response time and the mean service time respectively at station i. We can write:

$$E[T_i] = E[S_i] + E[S_i] \times E[N_i^*]. \tag{5.3.1}$$

The main difficulty is to find an expression for the magnitude of $E[N_i^*]$.

5.3.1. Closed networks with product form solution

If the network is closed and has a product form solution, the magnitude defined by relation (5.3.1) can easily be expressed.

Let $e_j, j = 1, \ldots, N$, be a solution of the system $e = ep$, which gives the mean number of visits to station j. We know (cf. Section 2.2) that if the network is closed, the values of $e_j, j = 1, \ldots, N$ are known only as approximate constants.

We can take as an example the solution determined by $e_i = 1$ where i is the queue which we are going to study. Calculations where the values $e_j, j = 1, \ldots, N$ are involved will be made with an approximate constant.

The time between two departures by the same customer from station i is:

$$\sum_{j=1}^{N} e_j E[T_j]/e_i. \tag{5.3.2}$$

Let K be the number of customers in the closed network. It has been shown that if the solution of the network is in product form (the joint probability is the product of the marginal probabilities), the mean number of customers found in queue i when entering it is the mean number in queue i in the stationary state when there is one less customer in the network. $E[N_i^*]$ is the mean number of customers in queue i in the stationary state, knowing that there are $K - 1$ customers in the network.

In the case of product form networks (BCMP type), we have the following relations:

$$E[T_i] = E[S_i] + E[S_i] \times E[N_i^*]. \tag{5.3.3}$$

Let $\lambda_i(K)$ be the arrival rate at station i knowing that there are K customers in the network. From relation (5.3.2) and Little's formula, we have:

$$\lambda_i(K) = \lambda(K)e_i$$

with

$$\lambda(K) = \frac{K}{\sum_{j=1}^{n} e_j E[T_j]}; \tag{5.3.4}$$

we have also:

$$E[N_i] = \lambda(K)e_i E[T_i]. \tag{5.3.5}$$

From relations (5.3.4) and (5.3.5) we obtain:

$$E[N_i] = \frac{Ke_i}{\displaystyle\sum_{j=1}^{n} e_j E[T_j]} E[T_i].$$

(5.3.6)

In expression (5.3.6) we see that the constant defining the values of e_i disappears.

From relations (5.3.3) and (5.3.6) we can obtain an exact solution by noting that $E[N_i] = 0$ for all i, if there are no customers in the system. In fact, one obtains $E[T_i]$ for $K = 1$ from equation (5.3.3). From relation (5.3.6), one obtains $E[N_i]$ for $K = 1$, then from (5.3.3), $E[T_i]$ for $K = 2$, etc...

The actual structure of the case examined is of little interest since the BCMP theorem applies and the solution is known in this way. However, the same method is applicable if there are different classes of customer or if the number of queues is increased; calculation of solutions by the BCMP theorem becomes impossible due to the very large number of states. As we are going to give, in the following, extensions to networks without a product form solution, we shall need only an approximate solution, allowing easy calculation of various interesting mean values.

The approximate solution which we propose involves assuming that the response time for a customer at station i is the same whether there are K or K − 1 customers in the entire network. This assumption will be better justified if there are many queues in the network or many customers in total.

Let $E[T_i^*]$ be the mean response time at station i given that there are K − 1 customers in the network. Our assumption is:

$$E[T_i] = E[T_i^*] = \frac{E[N_i]}{\lambda_i[K]}.$$

But from (5.6) we have:

$$E[T_i] = E[T_i^*] = \frac{\displaystyle\sum_{j=1}^{n} e_j E[T_j]}{e_i(K-1)} E[N_i^*].$$

By replacing $E[N_i^*]$ by its value in (5.1), we obtain:

$$E[T_i] = \frac{E[S_i]}{1 - E[S_i](K-1)e_i \Big/ \displaystyle\sum_{j=1}^{n} E[T_j]e_j}.$$

(5.3.7)

This is a non-linear relation between the $T_j, j = 1, \ldots, n$, which can be solved by simple numerical methods; for example, by using an iteration starting with finite non-zero values of $E[T_i]$. Since relation (5.3.7) has adequate properties of continuity and monotonicity, the solution found is therefore unique.

Comment Relation (5.3.3) is still valid if the service discipline is no longer

FIFO, but time division or last arrived, first served instead. In the case of an infinite number of servers, it is necessary to replace (5.3.3) by $E[T_i] = E[S_i]$.

5.3.2. Closed networks with different customer classes

R classes of customer, numbered from 1 to R, can use the network. We indicate the index of the class by an exponent r, $r = 1, \ldots, R$.

Relation (5.3.3) becomes:

$$E[T_i^r] = E[S_i^r] + E[S_i^r] \times \left[\sum_{\substack{r' \\ r' \neq r}} E[N_i^{r'}] + E[N_i^{r*}] \right], \qquad (5.3.8)$$

where $E[N_i^{r*}]$ is the mean number of customers in queue i in the stationary state since there are the same customers in the network less one customer of class r.

We are going to assume that customers cannot change classes in the network. This is not a restriction since one can always represent this case by increasing the number of classes. Let e_{ir} be the mean number (an approximate constant) of visits by a customer of class r to station i. We take $\Sigma e_{ir} = 1$. We again denote the total number of customers of class r in the network by K^r. We have:

$$\lambda^r(K) = \frac{K^r}{\sum_{j=1}^{n} e_{jr} E[T_j^r]}, \qquad r = 1, \ldots, R, \qquad (5.3.9)$$

and

$$E[N_i^r] = \lambda^r(K) e_{ir} E[T_j^r], \qquad r = 1, \ldots, R. \qquad (5.3.10)$$

Relation (5.3.6) can be replaced by the following:

$$E[N_i^r] = K^r \frac{e_{ir}}{\sum_{j=1}^{n} e_{jr} E[T_j^r]} E[T_i^r], \qquad r = 1, \ldots, R \qquad (5.3.11)$$

By noticing that $E[N_i^r] = 0$ if $K^r = 0$ for $r = 1, \ldots, R$, we can obtain the exact value of the mean number of customers in each queue and for each class by recurrence. The approximate method which we have indicated at the end of the preceding paragraph is always applicable. We describe an application of this method explicitly in Section 5.8.

5.3.3. Open networks

When the system is open, a unique solution exists for the system $e = q + ep$ which determines the number of visits by a customer of class r to station i (see Section 2.2): e_{ir}, $1 \leq i \leq n$ and $1 \leq r \leq R$.

If the solution of the open network is in product form, the number of customers in the queue at the time of entry is the same as that in the stationary state. This can be explained by the fact that if there is a product form, the stations behave as $M/M/1$ queues and we have seen in Section 1.2 that $p_k = q_k$ for all k. In the case of a first in, first out service discipline, in order to obtain the product form it is necessary that the service times are exponentially distributed and, if there are several classes of customer, that a unique service rate is defined at each station. When these conditions are not satisfied, there is no known result for the mean number of customers in a queue at the instant of arrival. In this case we assume that we are always concerned with the stationary state.

Let λ be the rate of arrival from the exterior.

The relations which govern the behaviour of queue i are as follows:

$$E[T_i^r] = E[S_i^r] + E[S_i^r] \times \sum_{r'=1}^{R} E[N_i^{r'}]$$

and

$$E[N_i^r] = \lambda e_{ir} E[T_i^r],$$

from which the following simple relation can be derived:

$$E[T_i^r] = E[S_i^r]\{1 + \lambda \sum_{r'=1}^{R} e_{ir'} E[T_i^{r'}]\}. \tag{5.3.12}$$

We obtain a linear Kramer system which is very simple to resolve.

Example

In the case of only two classes of customer, we obtain:

$$E[T_i^1] = \frac{\rho_{i1}\{E[S_i^1] - E[S_i^2]\} + E[S_i^2]}{1 - \rho_{i1} - \rho_{i2}},$$

$$E[T_i^2] = \frac{\rho_{i2}\{E[S_i^2] - E[S_i^1]\} + E[S_i^1]}{1 - \rho_{i1} - \rho_{i2}},$$

where

$$\rho_{ir} = \lambda e_{ir} \times E[S_i^r].$$

5.4 - AGGREGATION METHOD

5.4.1. Theory of the aggregation method

The principle of this method is to form groups of stations with the state variables of a complex system such that:

1. Interactions of variables within a group can be studied as if interactions with the exterior do not exist.

2. Interactions of groups can be studied without the need to consider interactions between variables within each group.

Systems which satisfy the necessary conditions for the aggregation technique to give exact results are called completely decomposable systems. They are formed from a certain number of sub-systems which operate independently of each other. When interactions between groups of variables are not zero but small compared with interactions within the groups, Courtois has stated certain conditions for which the aggregation of variables again gives satisfactory results; one then speaks of quasi-decomposable systems.

Let us consider a stochastic system described by the equations:

$$p(t + 1) = p(t)Q, \qquad (5.4.1)$$

where $p(t)$ is the line vector of state probabilities at time t and Q is the stochastic transition matrix of order $k \times k$. Let us also consider the system of the same dimension:

$$p^*(t + 1) = p^*(t)Q^* \qquad (5.4.2)$$

in which the matrix Q^* is composed of K square matrices on the principal diagonal, the other elements being zero. With these conditions, system (5.4.2) is completely decomposable.

Let us assume that a permutation of rows and columns of Q exists such that one can write:

$$Q = Q^* + \varepsilon C$$

in which C is a square matrix of which the sum of the elements of each row is zero; ε is a positive real number, small compared with the elements of Q^*. System (5.4.1) is then quasi-decomposable and can be studied by considering the system to be completely decomposable (5.4.2). Relations between the temporal operations of the two systems are given by Simon's and Ando's theorems for which we will give only a physical interpretation here.

In studying the dynamics of systems (5.4.1) and (5.4.2), if ε is sufficiently small, four periods separated by the instants T_1, T_2 and T_3 can be distinguished:

–a short-term transient period $(t < T_1)$ during which the two systems operate in an identical manner,
–a short-term equilibrium period $(T_1 < t < T_2)$; a stationary state is attained in each sub-system of (5.4.1) and (5.4.2),
–a long-term transient period $(T_2 < t < T_3)$; the quasi-decomposable system moves towards its global equilibrium state but the relative values of the variables obtained in each sub-system at equilibrium scarcely vary in the short term,
–a long-term equilibrium period $(t > T_3)$; system (5.4.1) attains its global equilibrium without the local equilibria being affected.

The decomposition method, therefore, reduces the study of quasi-decomposable systems to that of very small systems in two stages:

–calculation of the equilibrium state of K independent sub-systems,
–study of a new system of K dimensions of which the state variables are the aggregates obtained in the first stage.

The accuracy of the results for the global equilibrium is related to the quality of the approximation (order of magnitude of the constant ε) and has been analysed in detail by Courtois [16].

The decomposition can be made following a two level hierarchy of which one is occupied by the original system variables (microvariables) and the other by the aggregates (macrovariables). This hierarchy can be extended to more than two levels, each variable at a certain level being an aggregate of variables from the previous level.

The decomposition method applies, therefore, to Markovian queueing networks. In a queueing network an aggregation of variables is replaced by the concept of an *equivalent server* when exchanges between the aggregate and the outside world are made through a single server. This technique consists of replacing the part of the network which corresponds to the state variables of the aggregate by a unique queue. The two networks, before and after replacement, are equivalent if and only if:

– the probability distributions of the total number of customers in each of the two systems are identical,
– the probability distributions of the state variables which do not belong to the aggregate are identical.

The service time distribution of the equivalent server must be evaluated taking account of these constraints. The necessary parameters are obtained by studying the sub-network in the equilibrium state and applying the principles of the method of decomposition.

5.4.2. Example of the aggregation method

Study of a system with virtual memory

We have already studied a model of a system with virtual memory in Chapter 2. The system of queues was a network with a central server.

The system which we wish to examine is more comprehensive, we have represented it in Figure 5.3.

This model consists of a queue containing the 'process candidates' at the input of the central memory which is represented by a block of resources (the closed network of the central server). The characteristics of the block of resources can be obtained from a study of the closed network such as that carried out in Chapter 2.

The network of Figure 5.3 is a Markovian network, by assuming that arrivals from the exterior form a Poisson process and that all the service times are exponentially distributed and independent of each other. This system is Markovian but we cannot obtain a solution due to the interdependence of the three queues in the resource block which is expressed by $n_1 + n_2 + n_3 \leqslant K$ where n_1 is the number of customers in the central unit, n_2 the number of customers waiting for an input–output (or in service), n_3 the number of customers waiting for a secondary memory page (or in service) and K the maximum number of programs in the central memory.

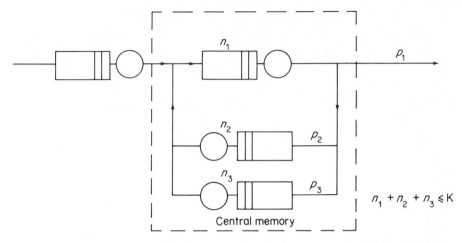

Figure 5.3 Equivalent model of system with virtual memory with $n_1 + n_2 + n_3 \leqslant K$

To study this system we are going to use the aggregation method. In fact the memory block can at first be studied alone (cf. Chapter 2) and replaced by an equivalent server of rate $\mu(m) = (1/C)A(m) \cdot p_1$ where C is the total mean computation time for a program in the central unit and $A(m)$ the utilization rate of the central server when the resource block contains m programs.

The new equivalent model is represented in Figure 5.4.

It only remains to study a system of two queues in series of the same type as those studied in Chapter 2.

5.5 - ISOLATION METHOD

5.5.1. Theory of the isolation method

The isolation method consists of subdividing the global system into L sub-systems and studying them separately.

To represent a system of queues, it is first necessary to determine a certain number of state parameters which enable definition of:

−the state of the global system,
−the state of the different sub-systems,
−the interfaces between each sub-system and the rest of the system.

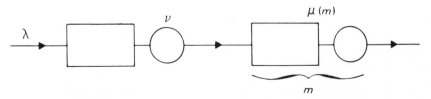

Figure 5.4 Equivalent model of system with virtual memory

Let $\mathscr{P}$ be the set of these parameters.

These parameters can represent the state probabilities of the system, in the usual sense of the term, but this is not a necessity. It is evident that such a representation of the total system is not generally rigorous since it necessarily leads to equations representing the global system which are difficult to resolve. The idea of the method is to model the interfaces by using simple methods to make them appear to be the most important parameters. At this stage, one has to make a choice which will originate in approximations already made or which will ultimately necessitate justification by other methods (simulation or measurement). Modelling of these interfaces allows the following to be known for each sub-system:

–the characteristics of the input streams (input interface function f_e),
–the characteristics of the service times (output interface function f_s).

Let $\mathscr{P} = \{p_i, 1 \leqslant i \leqslant J\}$ be the set of interface functions. Knowledge of a set of numerical values for these functions p_i allows all the characteristics of each sub-system to be known. Sub-system i has an equation of the form

$$p_i = f_i(p_j; j = 1, \ldots, J)$$

for $1 \leqslant i \leqslant J$. One now sees the necessity of choosing the decoupling in such a manner that each sub-system has a known solution.

To obtain the equations $p_i = f_i(p_j; j = 1, \ldots, J)$, one can use either an exact or an approximate method. Solution of the set of these J equations in J unknowns leads to a numerical solution of the global system.

When the system $(f_i, 1 \leqslant i \leqslant J)$ is formed, three problems arise:

(a) How to resolve the systems?
(b) Is the solution of the system equations unique?
(c) If it is unique, is it sufficiently close to the actual solution?

Solution of the system equations

In general, the equations do not form a linear system. One is therefore led to the use of numerical analysis methods. In most cases, one can easily obtain solutions with the required accuracy.

If, for example, all functions f_i are monotonic with respect to each of the parameters, this property can be exploited by using an iterative method. Initially the values of the parameters are fixed arbitrarily and the functions f_i are applied to these values as many times as are necessary to obtain the required precision.

Uniqueness of the solution

In general this is a difficult problem which must be examined for each particular case.

Validation

Validation of the results will be made by comparison with those obtained by simulation or by measurements made on a real system.

5.5.2. Example of the isolation method

In order better to understand this isolation method, we are going to develop a simple example taken from a study of the characteristics of a transport protocol between two nodes of a communication network called 'High-level Data Link Control' (HDLC). The HDLC protocol was chosen in 1975 as the standard protocol for communication between computers.

In the HDLC protocol, the frames (or packets) must conform to the following scheme:

Flag 01111110	Address 8 bits or 16 bits	Supervision 8 bits or 16 bits	Text	CRC* 16 bits	Flag 01111110

The length of the text is generally limited to 255 bytes. The packets are delimited by flags and the sequence 01111110 must not occur within a frame. To avoid this, the transmitter inserts a 0 element after every 11111 sequence. The receiver eliminates it. The address field must indicate the destination (or destinations) of the packet. The supervision field is reserved for correspondence between the receiver and transmitter. The following information is found there: the number of the particular frame and the number of the next awaited frame: n. This indicates that all frames numbered to $n-1$ have been received and discharged.

An important element of the HDLC protocol is the anticipation factor which determines the maximum number of packets which can be transmitted by the node without being discharged. The standard value for the HDLC protocol is 7 with an extended mode of 127.

We shall denote the anticipation factor of the HDLC protocol by N. Unlike the SW protocol, the parallelism of the processes requires us in this case to use two distinct levels to define the states of the transmitter and receiver: the software and the hardware. The software level corresponds to phenomena involving the central unit of the node, while the hardware level corresponds to phenomena associated principally with the coupling of lines. We shall assume that these two levels are independent.

The states used by the HDLC procedure are described by Figure 5.5.

In the following we shall assume that there are no line transmission errors. the transmission time of a packet S can be partitioned as follows:

$$S = \frac{1}{\mu} + \frac{1}{v}$$

(*) Cyclin Redundancy Checksum.

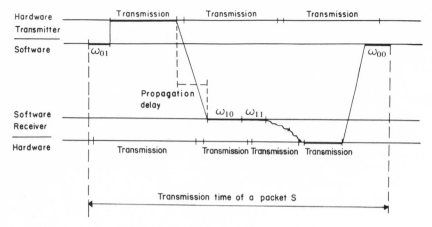

Figure 5.5 States of the HDLC protocol

where $1/\mu$ is the effective transmission time of the transmitter on the line and $1/v$ represents the other waiting times; propagation delay, supervision time, etc.

Parallelism operates at these time levels v^{-1} in such a way that a model of the HDLC protocol can be represented by servers in parallel. We have represented this in Figure 5.6.

It consists of a first station composed of a queue with a single server and a second station composed of N servers in parallel without a queue. When a customer finishes his service μ and finds the second station full, he restarts his service and this continues until he finds a place at the second station.

We shall consider each queue taken individually as a sub-system. A possible output interface for the first queue can be the following. We assume that the customer finishing his service sees the second station in its equilibrium state and so restarts his service with probability p (n customers in the second station). The equivalent service time which takes account of the times spent by a customer at the server has a mean of $[\mu(1 - p)]^{-1}$.

The input interface can be modelled by the set of customers who present

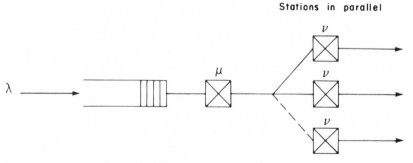

Figure 5.6 Model of the HDLC protocol

themselves to obtain a place in this queue. Such a customer is counted as many times as he enters the queue. In this case the arrival rate is $\lambda/(1-p)$.

It can be seen that the interface function is unique, it is the probability p of rejection. Studies of the first and second queues can be made separately but as a function of the parameter p. An explicit solution can be obtained by simultaneously solving the equations generated by the two sub-systems.

Let us put:

$$\tilde{\lambda} = \lambda/(1-p), \quad \tilde{\mu} = \mu(1-p), \quad \rho_1 = \frac{\lambda}{\tilde{\mu}}, \quad \rho_2 = \frac{\tilde{\lambda}}{v}.$$

The second queue is of M/G/C/C type which has the same solution as the M/M/C/C queue studied in exercise 3 of Chapter 1.

$$p = p(N) = \frac{\dfrac{\rho_2^N}{N!}}{1 + \rho_2 + \dfrac{\rho_2^2}{2!} + \cdots + \dfrac{\rho_2^N}{N!}}.$$

This is an equation with a single unknown p (notice that p occurs also in ρ_2) which can be solved numerically. Knowing the value of p, all the characteristics of queues 1 and 2 can be calculated.

As a numerical example, we shall calculate the maximum output as a function of the degree of anticipation of the protocol. The maximum output is obtained when there is always at least one packet ready to be transmitted in the first queue. In this case, the output of the first station (counting packets which will

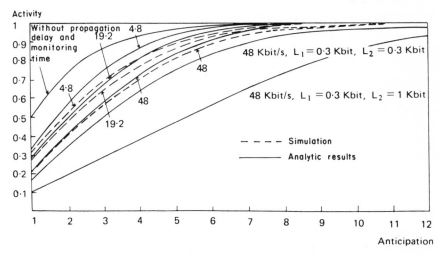

Figure 5.7 Normalized output as a function of the window for the HDLC protocol

L_2 = length of packets from transmitter to receiver

L_2 = length of packets from receiver to transmitter

be refused because the maximum anticipation is reached) is simply μ. It is sufficient to use the preceding formula giving p by taking $\rho_2 = \mu/v$. The maximum output of the system is then:

$$D = \mu(1 - p).$$

In Figure 5.7 we have sketched the activity of the line (that is $(1 - p)$) as a function of the degree of anticipation of the HDLC protocol for three distinct line speeds: 4.8 Kbits/second, 19.2 Kbits/second and 48 Kbits/second. The mean length of the packets is 1000 bits and their distribution is exponential. The various times which appear in Figure 5.5 have been taken equal to those obtained by field measurements on the French computer network CYCLADES.

5.6 - STUDY OF THE RESPONSE TIME OF A DATA TRANSMISSION NETWORK

A data transmission network is a computer system where numerous terminals and several computers are interconnected. The user addresses himself to the network and not to a particular machine. What happens in the network should be transparent to the users who demand only two things:

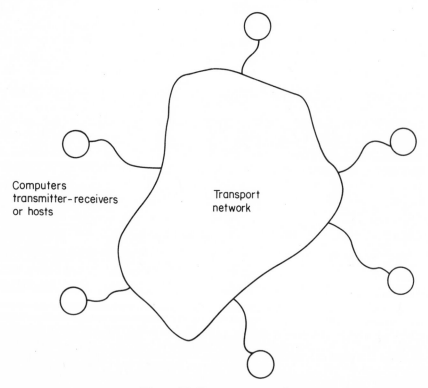

Figure 5.8 Computer network

1. that the requested tasks are correctly carried out
2. that the results arrive with maximum speed.

A response to the second question appears in this section. Firstly, the structure of a computer network will be examined.

Computer networks (Figure 5.8) can be separated into two sections: the actual transport network and the computer transmitters–receivers which we shall designate by the term global hosts.

The transport network is a distributed system consisting of computers which route blocks of data to their destinations. The size of these blocks or packets is usually between 20 and 250 bytes.

Two types of service are available for packet switching: transfer by virtual circuit or by datagram. The datagram service is a 'simple' packet transmission service more or less elaborated by the addition of routeing functions, stream control, etc. The virtual circuit service involves calling and releasing procedures in order to establish and close a connection. Buffers are reserved in the network nodes between two users.

Of course the two have advantages and disadvantages arising from their definition.

The pre-allocation of buffers involves an under-utilization of components, but allows the circulation of packets in the network to be more easily controlled. In the other case, allocation on demand, a better utilization of buffers is generally obtained but control of streams is more difficult.

Since the mathematical problems raised by reservation techniques are complicated, we shall calculate response times from a network datagram. We shall assume that the protocols which control the transfer of packets from node to node are of the 'stop and wait' type which we have already studied in Section 1.4. We shall also assume a fixed routeing, that is a packet can use only a well-specified route between a given source and destination.

Consider the communication network presented in Figure 5.9. It consists of four switching nodes numbered 1, 2, 3 and 4. Hosts H_1, H_2, H_3 and H_4 correspond to each node. Two different types of packet can enter from a host. The first category includes relatively short packets which arise in time division operation. The second category includes long packets which correspond to file transfers. These packets have practically the maximum length.

Let $S_{ij}, 1 \leqslant i \leqslant 4$ and $1 \leqslant j \leqslant 2$ be customers of class j leaving node i. The route of the packets in the network is determined by the shortest path. When there are two possible paths of the same length, the route adopted is that indicated in Figure 5.9.

There is no change of class within the network. Consequently, a long packet leaving a certain node cannot be transformed into a short packet leaving another node. To define completely the model, it is necessary to know the destination probability of packets of each class. Table 5.1 gives all the information on this subject.

In summary, we have a network which can be modelled by a system of queues

136

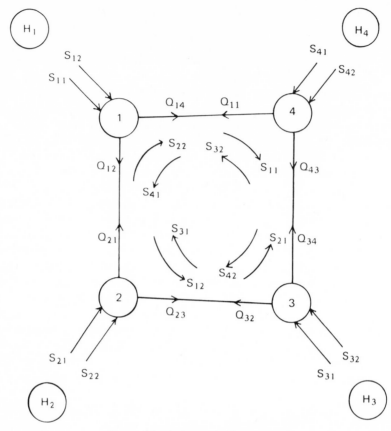

Figure 5.9 Packet switching network

Table 5.1 - **Characteristics of classes of customers**

Class	Routeing				Law of arrival in the network	Law of arrival at nodes
	1	2	3	4		
S_{11}		0.25	0.375	0.375	M(C)	M(0.1)
S_{12}		0.335	0.335	0.33	H_2(D)	D(0.5)
S_{21}	0.33		0.335	0.335	M(C)	M(0.1)
S_{22}	0.335		0.33	0.335	H_2(D)	D(0.5)
S_{31}	0.45	0.22		0.33	E_2(C)	M(0.2)
S_{32}	0.335	0.33		0.335	M(D)	E_5(0.4)
S_{41}	0.375	0.375	0.25		E_2(C)	M(0.1)
S_{42}	0.33	0.22	0.45		H_3(D)	D(0.5)

Table 5.2 - Mean number of customers and response times for the network represented in Figure 5.6

		Q_{14}		Q_{12}		Q_{21}		Q_{23}		Q_{32}		Q_{34}		Q_{43}		Q_{41}	
		L	T	L	T	L	T	L	T	L	T	L	T	L	T	L	T
C = 0.6 D = 3	*Sim	0.309	0.213	0.284	0.225	0.453	0.298	0.294	0.221	0.380	0.297	0.352	0.262	0.281	0.223	0.292	0.199
	*TD	0.309	0.209	0.275	0.217	0.456	0.301	0.283	0.214	0.429	0.334	0.347	0.259	0.272	0.216	0.292	0.198
	*Dif.	0.313	0.212	0.277	0.219	0.451	0.297	0.286	0.216	0.416	0.324	0.347	0.259	0.275	0.218	0.297	0.201
C = 0.6 D = 1	Sim	0.798	0.417	0.660	0.401	1.122	0.582	0.777	0.445	0.721	0.439	0.693	0.389	0.912	0.535	0.745	0.387
	TD	0.843	0.440	0.783	0.457	1.159	0.591	0.793	0.449	0.854	0.518	0.774	0.433	0.767	0.451	0.751	0.390
	Dif.	0.798	0.417	0.728	0.425	1.040	0.530	0.743	0.421	0.779	0.472	0.731	0.409	0.814	0.420	0.725	0.377
C = 0.6 D = 0.8	Sim	1.044	0.505	1.145	0.623	1.651	0.768	1.148	0.578	0.910	0.514	0.926	0.464	1.276	0.697	0.973	0.463
	TD	1.177	0.565	1.096	0.583	1.636	0.768	1.108	0.573	1.088	0.609	1.013	0.518	1.069	0.572	1.020	0.487
	Dif.	1.074	0.516	0.979	0.521	1.606	0.766	1.111	0.579	0.965	0.541	0.993	0.478	0.957	0.513	0.958	0.457

*Sim. = Simulation; TD = Time Division; Dif. = Diffusion.

containing eight queues denoted by Q_{ij}, i being the origin node and j the destination node. We assume that these queues have an infinite capacity; packets are never rejected following an overflow. This last assumption is the principal approximation of the model compared with reality. The arrival rate of the classes of packets and also the distribution of the arrival process are specified in Table 5.1. In particular, long packets arrive in sections; a file transfer generally requires more than one packet. The probability distribution of the length of these packets is constant or follows an Erlang-5 distribution: the size of the packets is fixed in practice. In contrast, short packets generally arrive alone and irregularly, their length is exponentially distributed. The type of traffic which we have described is characteristic of applications in computer networks.

The mathematical technique which will be used to solve this sytem of queues is that described in Section 5.2. In fact, the BCMP theorem is not applicable since on the one hand the external arrivals are not exponentially distributed and on the other hand the service times of various classes of customer are neither exponentially distributed nor identical at each station of the network. In contrast, if we could assume that the service discipline is no longer first come, first served but time division, the second condition for BCMP networks would be satisfied. If we also assume that the interarrivals are exponentially distributed, the BCMP theorem would be perfectly applicable.

In Table 5.2 we have collected the results of the following three methods:

1. results obtained from a simulation of the proposed model,
2. results obtained by assuming a time division service discipline and exponentially distributed arrivals from the exterior with the same mean as that given in Table 5.1,
3. results obtained by the approximate general method given in Section 3.3 using different classes of customer.

We have introduced simulation since there is no available method which gives exact results. In the following tables we give only the centres of the confidence intervals (95% confidence) for these simulation results and they allow only an estimate of the validity of the two other methods.

From these results, it is clear that the method described in Section 5.2 is very well suited to the type of problem under consideration. The approximation which involves assuming a time division discipline also gives very good results when the load rate is small.

5.7 - APPLICATION TO THE STUDY OF THE CHARACTERISTICS OF AN INTERACTIVE COMPUTER

We again use a simplified form of the model which was introduced in Chapter 2 and which is represented in Figure 5.10.

Unlike the application of Section 2.7, several classes of customer are allowed. These classes of customer represent programs which do not all have the same behaviour with respect to the central processing unit and the secondary memory.

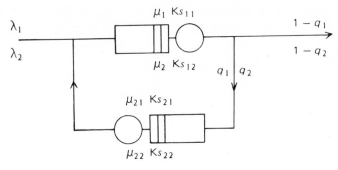

Figure 5.10 Model of an interactive computer

We shall assume that class 1 makes numerous demands on the secondary memory whereas class 2 represents programs which consume time on the central processing unit. The various parameters which we have chosen are given in Table 5.3.

We have assumed in particular that the time intervals between arrivals from the exterior are exponentially distributed

$$(Ka_1 = Ka_2 = 1).$$

The corresponding results for various values of the parameters are given in

Table 5.3 - **Mean number of customers (L_1 and L_2) and response time (T_1 and T_2) of the network represented in Figure 5.10.**

	μ_{11}	μ_{12}	Ks_{11}	Ks_{12}	μ_{21}	μ_{22}	Ks_{21}	Ks_{22}	L_1	L_2	T_1	T_2
				$\lambda_1=0.5$	$\lambda_2=0.25$	$Ka_1=1$	$Ka_2=1$	$q_1=0.5$	$q_2=0.2$			
Simulation	2	1	1	0	2	1	2	1	3.62	0.49	2.79	0.89
Time division	„	„	„	„	„	„	„	„	4.33	0.45	3.30	0.81
Diffusion	„	„	„	„	„	„	„	„	3.66	0.50	2.79	0.90
Simulation	„	„	0.2	„	„	„	„	„	3.15	0.45	2.43	0.81
Time division	„	„	„	„	„	„	„	„	4.33	0.45	3.30	0.81
Diffusion	„	„	„	„	„	„	„	„	2.91	0.49	2.22	0.88
Simulation	„	2	1	1	1		1	„	1.92	1.24	1.46	2.18
Time division	„	„	„	„	„	„	„	„	1.90	1.29	1.45	2.28
Diffusion	„	„	„	„	„	„	„	„	1.90	1.29	1.45	2.28
Simulation	„	„	3	0	2	„	2	1	2.35	0.51	1.81	0.92
Time division									1.91	0.45	1.45	0.81
Diffusion									2.77	0.53	2.11	0.94
Simulation	„	„	„	„	„	„	1	0.2	2.37	0.50	1.84	0.90
Time division									1.91	0.45	1.45	0.81
Diffusion									2.77	0.50	2.11	0.90

Table 5.3. They concern the mean lengths of the two queues of the model. These results have been obtained by three different methods as in the preceding paragraph:

1. by the general method given in Section 5.2 using Kingman's formula,
2. by simulation for which we give the centre of the confidence interval (at 95%),
3. by the Baskett, Chandy, Muntz and Palacios theorem assuming that the service disciplines are time division.

The values found in these different cases merely confirm the results of the preceding section; the approximate general method allows results whose quality is equivalent to those of a simulation to be obtained at very low cost.

5.8 - PERFORMANCE STUDY OF A COMPUTER NETWORK WITH VIRTUAL CIRCUITS

Computer networks can be separated into two major categories:

–datagrams in which packets enter the network as they arrive,
–virtual circuits in which a packet can enter the network only if a virtual circuit is assigned to it.

The network which we have studied in Section 5.6 was of the datagram type. For the second category it is generally necessary to have a calling packet which defines the circuit (it will be formed only if the system resources are sufficient) and a closing packet which cancels the virtual circuit. The virtual circuit is, therefore, a path in the network which will be followed by the packets allocated to this circuit and which activates a certain number of resources in the network such as buffers in the switching nodes.

The difficulty in studying virtual circuit networks is the dependence of inputs to the network on already allocated interior resources. However, using the mean value technique, given in Section 5.3, we can study this type of network and deduce some of its characteristics.

A virtual circuit between two hosts or terminal equipments is represented in Figure 5.11.

Each packet transmitted in a virtual circuit is numbered cyclically modulo

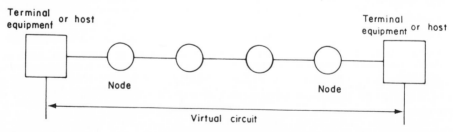

Figure 5.11 Virtual circuit

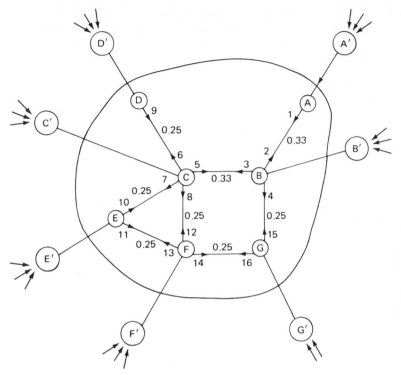

Figure 5.12 The network analysed

L + 1 (integer). Anticipation of up to L packets is allowed, that is up to L packets can be sent without receiving cancellations.

One of the major difficulties encountered with this type of network is control of the output which is realized by varying the values of anticipation and number of virtual channels. For example, can one double the output by doubling the number of resources and hence virtual channels? A study will show us that this is impossible.

As an example, we shall consider the network of seven nodes described in Figure 5.12. Nodes A to G consist of mini-computers which store and switch the packets.

Nodes A' and G' represent the terminal equipments. Routeing of the packets (that is establishment of the virtual circuits) is assumed to be fixed and is summarized in Table 5.4. For routes which are not indicated in Table 5.4 the shortest is chosen. We assume that the lengths of packets are exponentially distributed and that these arrive at the terminal equipment following a Poisson process. We shall also assume that packets arriving from the exterior have an equal probability of going to one of the six other hosts.

The packet switching network represented in Figure 5.12 can be modelled by a system of 16 internal queues and 7 input queues (one for each host). We

Table 5.4 - **Complementary routes**

Source	Destination	Via
A′	F′	C
B′	F′	C
C′	G′	F
D′	G′	F
G′	C′, D′	B
F′	A′, B′	C

assume that node–host output queues are non-existent, since they correspond to much faster lines.

We are also going to assume that the 42 possible host–host paths each possess a virtual circuit, all allowing the same anticipation. A virtual path having an anticipation window of L can be replaced by L virtual circuits with a window of 1. In the example, all lines have the same transfer capacity and the mean number of packets which can be transmitted in unit time is taken equal to $10 (\mu = 10)$. The 0.1 second service time takes account of various control and management functions associated with the transmission procedure.

In the first case, we shall study the response time of four virtual circuits by varying the length of the anticipation window. We shall consider the following connections:

connection 1: host A′ to host D′,
connection 2: host A′ to host F′,
connection 3: host D′ to host G′,
connection 4: host E′ to host G′.

The method of solution is that proposed in Section 5.3. We have 42 L classes of customer. With L = 10 that represents 420 classes of customer which cannot be solved by applying the BCMP theorem because the number of states is much too large. (We have assumed a FIFO service discipline, exponential service times and the same mean for each class at each station; the BCMP theorem is therefore theoretically applicable.)

In the example studied we have simply $\rho_{ir} = 1$ if customer class r passes through queue i. We shall assume that the load on the network is constant and that $\lambda L = 1$ where λ is the arrival rate of customers of class r.

From Figure 5.12 we obtain the following relations:

$$E[T_i^r] = E[S_i^r]\left\{1 + \lambda \sum_{r'=1}^{R} E[T_i^r]\right\}, \quad r = 1,\ldots,42L.$$

We obtain a system of 42 L equations with the same number of unknowns. A set of solutions is represented in Figure 5.13. It should be noted that the response time decreases as the maximum anticipation increases, but it decreases non-linearly. Above an anticipation of 4, the response time scarcely varies.

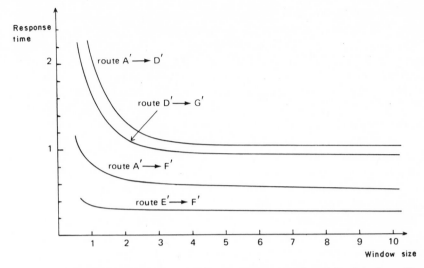

Figure 5.13 Response time as a function of window size when $L\lambda = 1$ and $\mu = 10$

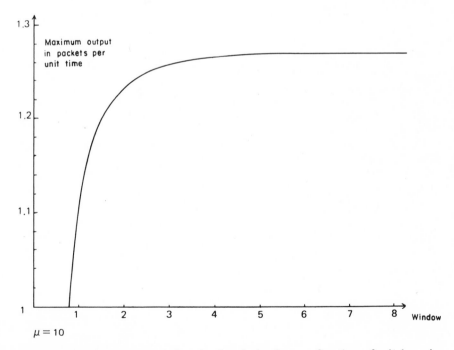

Figure 5.14 Maximum output of each virtual circuit as a function of window size (identical for each circuit)

In Figure 5.14 we give the maximum output for each virtual circuit as a function of the maximum anticipation. The bottleneck of the network is queue 3. The virtual circuit $A' \rightarrow D'$ is that which limits the output. The result which we obtain in Figure 5.14 for the increase of the function could be foreseen. In contrast, the important point is that above a window of 4 the output scarcely increases. This follows simply from the fact that increasing the anticipation window above a certain limit does not increase the output. In this example it can be clearly seen that the output cannot be uniquely regulated by adjusting the size of the anticipation window. Furthermore, in the preceding example, if one assumes that a virtual circuit with a window length of L needs to reserve L buffers at each node, the number of buffers increases rapidly. For example for queue 3 at the output of node B, there are 8 virtual circuits and hence 8 L buffers will be necessary for allocation of all the virtual circuits.

This performance study shows that there is an optimum for the size of the anticipation window. To increase this optimal size requires more system resources in terms of memory capacity without which no increase in output is noticeable. Below this optimal size the output decreases very rapidly.

BIBLIOGRAPHY

In the introductory section, we proposed a classification of analysis methods for queueing systems which is published in [1]. In the same book, one can find details of the two other major approximation methods not treated here, that is numerical methods (cf. also [2][3]) and iterative methods (cf. also [4] and [5]).

The decomposition method has been introduced in [6] for the general case of a network of ./G/1 queues, then in [7] for the case where several classes of customer can coexist.

Mean value analysis appeared in papers [8] and [9]. The property used, showing that the number of customers at the input of a queue is the mean number in the stationary state, has been demonstrated by Sevcik and Mitrani [10] and also by Lavenberg and Reiser [11]. Extensions of this method are considered in [12] and [13].

The aggregation method has been particularly developed by Courtois in [14][15][16]. Examples of applications can be found in [17][18][19][20].

The isolation method described in Section 5.5 arises from paper [21].

1. Gelenbe, E., Labetoulle, J., Marie, R., Metiver, M., Pujolle, G., and Stewart, W. (1980). *Réseaux de files d'attente*, Éditions Hommes et Techniques.
2. Stewart, W. J. (1977). A new approach to the numerical analysis of Markovian models, In *Computer Performance*, Editors K. M. Chandy and M. Reiser, North Holland publishers.
3. Stewart, W. J. (1978). A comparison of numérical techniques in Markov modelling, *CACM*, **21**, 144–152.
4. Chandy, K. M., Herzog, U., and Woo, L. (1975). Approximate analysis of general queueing networks, *IBM J. Res. Dev.*, 43–49.
5. Marie, R. (1978). Méthodes itératives de résolution de modèles mathématiques de systèmes informatiques, *RAIRO Informatique*, **12**,
6. Gelenbe, E., and Pujolle, G. (1976). The behaviour of a single queue in a general queueing network, *Acto Informatica*, **7**, 123–136.
7. Gelenbe, E., and Pujolle, G., (1977). A diffusion model for multiple class queueing

networks. *Proc. of the 3rd Internat. Symp. on Modelling and Performance Evaluation of Computer Systems, Bonn.*

8. Reiser, M. (1979). *Mean value analysis of queueing networks, a new look at an old problem,* – IBM Research Report RC, 7228.
9. Labetoulle, J., and Pujolle, G. (1981). A study of flows through virtual circuits computer networks. *Computer Networks*, **5**, 119–126.
10. Sevcik, K., and Mitrani, I. (1979). The distribution of queueing network states at input and output instants. *Proc. Int. Symp. Performance Computer Systems, Vienna.*
11. Lavenberg, S., and Reiser, M. (1979). *The state seen by an arriving customer in closed multiple chain queueing networks,* IBM Research Report Yorktown Heights.
12. Reiser, M. (1980). A queueing network analysis of computer communication networks with window flow control, *IEEE Trans. Com.*
13. Bard, Y. (1979). Some extensions to multiclass queueing network analysis, *Proc. Performance of Computer Systems,* Vienna.
14. Courtois, P. J. (1971). *On the near complete decomposability of network of queues and of stochastic model of multiprogramming systems,* – Carnegie-Mellon University Research Report, Computer Science Dept.
15. Courtois, P. J. (1975). Decomposability, instabilities and saturation in multiprogramming systems, *CACM*, **18**, 371–377.
16. Courtois, P. J. (1977). *Decomposability Queueing and Computer Science Applications,* ACM monograph series, Academic Press, New York.
17. Brandwajn, A., Buzen, J., Gelenbe, E., and Potier, D. (1974). A model of performance for virtual memory systems, *Proc. ACM – Signetics Symposium,* Montreal.
18. Badel, M., Gelenbe, E., Leroudier, J., and Potier, D. (1975). Adaptive optimization of a virtual memory system, *Proc. IEEE Special Issue Interactive Systems.*
19. Brandwajn, A. (1974). A model of time-sharing virtual memory system solved using equivalence and decomposition method, *Acta Informatica*, **4**, 11–47.
20. Gelenbe, E., and Kurinckx, A. (1978). Random injection control of multiprogramming in virtual memory, *IEEE Trans. Software Engineering*, **4**.
21. Labetoulle, J., and Pujolle, G. (1980). Network of queues, *IEEE Trans. Software Engineering*, **6**, 373–381.

CHAPTER 6

Flows in Networks

6.1 - INTRODUCTION

The aim of this chapter is to explain and to study thoroughly a certain number of important results for flows in the interior of queueing networks. In fact, study of the networks may have characterization of the arrival and departure processes of the stations as a prerequisite. An approximate method of this type has been proposed in the preceding chapter. After specifying the terminology used in a general network, we give several results for the output process of a simple queue, by assuming that the arrival process is known. Flows in Jackson networks are then studied and extensions are considered.

Let us recall rapidly the general context in which the results of this chapter occur. Consider a general network which has N queues such as that represented in Figure 6.1. The probabilities p_{ij} of going from a queue i to a queue j are determined by a Markov chain of order 1. Station 0 is a fictitious station representing entrance to and exit from the system such that p_{0i} is the probability

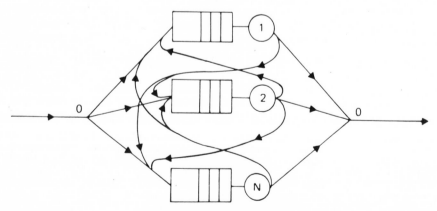

Figure 6.1 General network

146

that a customer arriving from the exterior goes to queue i and p_{j0} is the probability that a customer finishing his service at station j leaves the system. Let $k = (k_1, k_2, \ldots, k_N)$ be the number of customers in queues $1, 2, \ldots, N$ respectively. The behaviour of the network is totally defined by the values of the $p(k, t)$: the probabilities of being in state $(k_1, k_2, \ldots, k_N)$ at time t which we shall call the conjoint probabilities. We shall use the same notation for the marginal probabilities $p(k_i, t)$ of having k_i customers in queue i at time t. The context avoids confusion between conjoint and marginal probabilities.

A queueing network will be said to be in equilibrium if a stationary state exists. In this case, we shall eliminate time t from the joint and marginal probabilities.

By definition we say that *a network is of product form* if

$$p(k) = \prod_{i=1}^{N} p(k_i).$$

We shall use the following terminology; inputs to and outputs from a queue consist of customers entering and effectively leaving the queue; arrivals and departures consist of flow(s) taken separately coming from other stations or the exterior and leaving for other stations or the exterior. Inputs are a superposition of arrivals, and outputs give rise to departures if customers are neither rejected nor lost at the input or the interior of a queue.

6.2 - OUTPUT PROCESS OF A QUEUE

Many publications are concerned with output properties. We summarize the principal stages before considering the conditions required for the departure process to be a Poisson process. In Section 6.3 we record proofs of the principal results which we shall give here.

In 1956 Burke [3] showed that the departure process of an M/M/C queue was Poisson. He established the proof from a demonstration of the independence of an interval between two outputs and the state of the system at the end of this interval. At almost the same time Reich [4] showed the same result by using the reversibility of a stationary stochastic process. In 1959 Finch [5] studied the M/GI/1/m queue with a twice differentiable service time distribution. He showed that, in the equilibrium state, the only case where the inter-outputs form a Poisson process independent of the input process is obtained for GI = M and $m = \infty$. Finally, two articles by Disney and Cherry [6] and Disney, Farrel and De Morais [7] demonstrate the following theorem:

Theorem

An M/GI/1/m queue has output instants which form a renewal process if and only if one of the following conditions is satisfied:

1. the service times are all zero with probability 1,
2. $m = 1$,

3. $m = 2$ and the service times are constant (GI = 0),
4. $m = \infty$ and the service times are exponential (GI = M).

In these four cases, the respective probability distributions of the inter-outputs are the following:

1. the same as that of the input process,
2. the convolution of that of the input process and the service process,
3. a sum of convolutions,
4. the same as that of the input process.

Furthermore, it has been shown by Daley [8] that among GI/M/1 systems, the only one which has a renewal output process is again the M/M/1 system. More recently, Laslett [9] has shown that any GI/M/1/m system with finite m does not have a renewal output process.

We know, therefore, that the M/M/1/∞ system has a Poisson output process independent of the input process.

From this result, we find a first set of product form networks; they are networks having only type M/M/1 queues, that is the networks have type ./M/1/∞ queues where there is never a possibility that a customer passes the same station twice and where arrivals from the exterior are Poisson. Such a network is represented in Figure 6.2; it is a particular case of Jackson networks.

It should be noted that in such networks all the flows are independent Poisson processes.

The second type of network which we shall study consists of those which satisfy the local balance equations; the stream of customers arriving at a station containing n customers is equal to a stream of customers leaving this same station and leaving n customers there. These local balance equations involve only the state of a single queue at a time. This proves the mutual independence

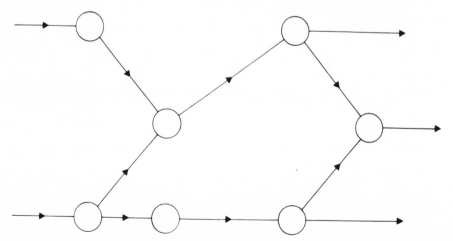

Figure 6.2 Queueing network with Poisson streams

of the queues. Hence a network which satisfies the local balance equations has a product form solution.

Most networks which satisfy the local balance equations are given in the paper by Baskett *et al.* [12] and Jackson networks are a sub-set of these. If only the case of first in, first out discipline is considered, the conditions for a product form are:

–the service times follow exponential laws,
–the arrival processes from the exterior are Poisson,
–the service time distribution of a solution is the same for all customers (even if there are several 'classes' of customer).

For three other service disciplines (last in, first out with absolute priority, time division and infinite number of servers), the service times can have rational Laplace–Stieltjes transform distributions.* For these cases, several classes of customer with different service time laws can be admitted to the same station.

Before tackling an explicit calculation of flows in Jackson networks, we are going to obtain several preliminary results by reversibility methods, due above all to Kelly [13], of which most have already been mentioned in this section.

6.3 - REVERSIBILITY AND QUASI-REVERSIBILITY

A stochastic process $N(t)$ is reversible if for any sequence of times $t_1, \ldots, t_n$, and for all times T, the joint distribution of the variable $(N(t_1), \ldots, N(t_n))$ is identical to that of

$$(N(T - t_1), N(T - t_2), \ldots, N(T - t_n)).$$

Any process may be reversible, which is not the case as will be seen for a quasi-reversible process. When the stochastic process $N(t)$ is Markovian and stationary, a necessary and sufficient condition for reversibility is that the solution $p(k)$ in the stationary state satisfies the equations which we shall call detailed balance equations. If $d(k, k')$ is the transition rate from state k to state k', they are written:

$$p(k)d(k, k') = p(k')d(k', k).$$

Before demonstrating this property we give two more explicit examples.

Examples

1. Let us take the case of the M/M/1 queue with a service rate depending on a function of the number of customers: $\mu(k)$. We have $d(k, k') = 0$ except when $k' = k + 1$ or $k - 1$. In this case the detailed balance equations are:

$$\lambda p(k) = \mu(k + 1)p(k + 1),$$

for each state k.

(*)And even general independent distributions.

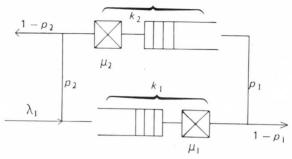

Figure 6.3 Jackson network with two queues

One again finds the local balance equations which we met in Chapter 3. In contrast, in the second example which we shall give, the local balances and detailed balances are not identical.

2. Consider the Jackson network represented in Figure 6.3.

We are going to treat only the case where k_1 and k_2 are strictly positive to show the three types of balance equation.

The global balance equation can be written:

$$
\begin{aligned}
(\lambda_1 + \mu_1 + \mu_2)p(k_1, k_2) = {} & \lambda_1 p(k_1 - 1, k_2) \\
& + \mu_1 p_1 p(k_1 + 1, k_2 - 1) \\
& + \mu_1(1 - p_1)p(k_1 + 1, k_2) \\
& + \mu_2 p_2 p(k_1 - 1, k_2 + 1) \\
& + \mu_2(1 - p_2)p(k_1, k_2 + 1).
\end{aligned}
$$

The three local balance equations are written in the following manner:

$$\lambda_1 p(k_1, k_2) = \mu_1(1 - p_1)p(k_1 + 1, k_2) + \mu_2(1 - p_2)p(k_1, k_2 + 1)^*,$$
$$\mu_1 p(k_1, k_2) = \lambda_1 p(k_1 - 1, k_2) + \mu_2 p_2 p(k_1 - 1, k_2 + 1).$$
$$\mu_2 p(k_1, k_2) = \mu_1 p_1 p(k_1 + 1, k_2 - 1).$$

As we have seen in Chapter 2, the local balance equations are satisfied in Jackson networks. This can be verified in the example treated here since the solution is simple:

$$p(k_1, k_2) = (1 - \rho_1)(1 - \rho_2)\rho_1^{k_1}\rho_2^{k_2},$$

with

$$\rho_1 = \frac{\lambda_1 \rho_1}{\mu_1} \quad \text{and} \quad \rho_2 = \frac{\lambda_1 \rho_2}{\mu_2},$$

from which

$$\rho_1 = \frac{1}{1 - p_1 p_2} \quad \text{and} \quad \rho_2 = \frac{p_1}{1 - p_1 p_2}.$$

Finally, the detailed balance equations are written in the following manner:

*N.B. This first equation corresponds to the local balance equation of the external station (number 0)

$$\mu_1 p_1 p(k_1, k_2) = \mu_2 p_2 p(k_1 - 1, k_2 + 1),$$
$$\lambda_1 p(k_1, k_2) = \mu_1 (1 - p_1) p(k_1 + 1, k_2),$$
$$\mu_2 p_2 p(k_1, k_2) = \mu_1 p_1 p(k_1 + 1, k_2 - 1),$$
$$\mu_1 (1 - p_1) p(k_1, k_2) = \lambda_1 p(k_1 - 1, k_2).$$

One immediately notices that the detailed balance equations are not satisfied in the general case. From the theorem which we have stated and will demonstrate, the Jackson network of Figure 6.3 is not a reversible network.

Let us now show that the stationary Markovian process $N(t)$ is reversible if and only if the detailed balance equations are satisfied.

The transition rate is defined by:

$$d(k, k') = \lim_{\tau \to 0} \frac{P\{N(t + \tau) = k' \mid N(t) = k\}}{\tau}.$$

If the process is reversible we have:

$$p(k)d(k, k') = \lim_{\tau \to 0} P\{N(t + \tau) = k', N(t) = k\}/\tau,$$
$$= \lim_{\tau \to 0} P\{N(T - (t + \tau)) = k', N(T - t) = k\}/\tau$$

(by putting $T = 2t + \tau$)

$$p(k)d(k, k') = \lim_{\tau \to 0} P\{N(t) = k', N(t + \tau) = k\}/\tau,$$
$$= p(k')d(k', k).$$

Conversely, consider a series of instants $t_1, \ldots, t_n$ of successive jumps of the process $N(t)$ and denote by $\alpha k_i(t)$ the probability that the process remains in state k_i during time t.

$$P\{N(t_1) = k_1, N(t_2) = k_2, \ldots, N(t_n) = k_n\}$$
$$= p(k_1)\alpha_{k_1}(t_2 - t_1)d(k_1, k_2)\alpha_{k_2}(t_3 - t_2)d(k_2, k_3)\ldots d(k_{n-1}, k_n).$$

By using the detailed balance equations the preceding expression becomes:

$$P\{N(t_1) = k_1, N(t_2) = k_2, \ldots, N(t_n) = k_n\}$$
$$= \alpha_{k_1}(t_2 - t_1)d(k_2, k_1)p(k_2)\alpha_{k_2}(t_3 - t_2)\ldots,$$
$$= \alpha_{k_1}(t_2 - t_1)d(k_2, k_1)\alpha_{k_2}(t_3 - t_2)\ldots d(k_n, k_{n-1})p(k_n),$$
$$= P\{N(-t_n) = k_n, N(-t_{n-1}) = k_{n-1}, \ldots, N(-t_1) = k_1\},$$

this is due to the fact that the process is Markovian. Finally, since the process is stationary, the reversibility condition is satisfied and one has:

$$= P\{N(T - t_1) = k_1, \quad N(T - t_2) = k_2, \ldots N(T - t_n) = k_n\}.$$

Reconsider the example of the M/M/1 queue. The number of customers in the queue is a stationary Markovian process if $\lambda < \mu$. As the detailed balance equations are satisfied, the queue is reversible. The instants of arrival in the

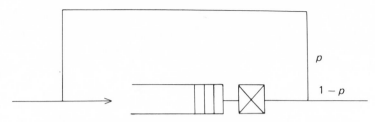

Figure 6.4 M/M/1 queue with feedback

queue correspond inversely to the times of output instants from the queue of the inverse process. Since the input process is Poisson, the output process is also. Let t_0 be any instant. As the arrival process in the queue after time t_0 does not depend on the number of customers in the queue up to time t_0, by examining the inverse process it can be seen that the state of the queue at time t_0 does not depend on the departure process up to time t_0.

Study of the M/M/1 queue with feedback

We shall study the M/M/1 queue with feedback represented in Figure 6.4 by the method described previously.

The process N(t) describing the number of customers in the queue is a Markov process. The detailed balance equations (equivalent to the local balance equations) are satisfied so the system is reversible. The input process of the queue is deduced from N(t) when this jumps upwards from 1. By considering inverse time N(−t) has the same distribution as N(t) since the process is reversible, the input process is identical to that of the outputs for the queue N(t). This shows that the output process is equivalent to the inverse of the input process.

Queues in tandem

Consider a series of queues (Figure 6.5) all having exponentially distributed service times.

If the input process of the system is Poisson, the output process of the first queue is also Poisson and for any instant t_0, it does not depend on the state of the queue at time t_0. This proves that the state of the first queue $p_1(n, t_0)$ is independent of the state of the following queues; the product form and the Poisson character of all the flows of this network are therefore confirmed.

We have seen from a particular example that in general a Jackson network is not reversible, in contrast we are going to introduce a new less strong property

Figure 6.5 Queues in tandem

which will be satisfied by Jackson networks. Let us call a set of queues a system. A system is a black box having inputs and outputs.

Definition

A system is quasi-reversible if its state is described by a stationary Markovian process $N(t)$ such that $N(t_0)$, for all times t_0, is independent:

−of the input process after t_0,
−of the output process before t_0.

Theorem

If a system is quasi-reversible, then the input and output processes are independent Poisson processes.

Let $d(k) = \sum_{k' \in A(k)} d(k, k')$, where $A(k)$ is the set of system states which possess one element more than state k. Since $N(t)$ is a Markovian process, the state at time $t_0 + t$ depends only on the state at time t. Furthermore, from the assumption, the arrival process after t_0 does not depend on $N(t_0)$. This implies that the arrival process is random, that is it consists of a Poisson process.

A departure from the system corresponds to an arrival in the inverse system. Hence a departure is a change from state k to state $k' \in A(k)$ in the inverse system. Since the system is quasi-reversible, the inverse system is also quasi-reversible. The same reasoning as that followed previously shows that arrivals for the inverse system form a Poisson process and hence the output process of the system is also Poisson.

Theorem

A system is quasi-reversible if the associated process is Markovian and satisfies the local balance equations.

Let $d'(k, k')$ be the transition rates of the inverse process. From what we have seen:

$$d(k) = \sum_k d'(k, k').$$

Furthermore we have:

$$d'(k, k') = \frac{p(k')}{p(k)} d(k', k)$$

where again:

$$p(k)d'(k, k') = p(k')d(k', k);$$

by summing over $k' \in A(k)$, one obtains:

$$p(k) \sum_{k' \in A(k)} d(k, k') = \sum_{k' \in A(k)} p(k')d(k', k)$$

which represents the local balance equations.

In Jackson networks as well as networks which satisfy the BCMP theorem, the local balance equations are satisfied. These networks are quasi-reversible and in particular the output processes are Poisson processes. In the following section, we shall study different flows in the interior of Jackson networks.

6.4 - FLOWS IN JACKSON NETWORKS

6.4.1. Flow in the M/M/1 system with feedback

To show the interest of the problem, we shall start with a very simple example; the case of an M/M/1 queue with feedback which we have represented in Figure 6.6.

In this queue, at the end of a service the customer departs with probability q, or returns to the input (feedback) with probability $p = 1 - q$, independently of all other events.

Let $0 < a_1 < a_2 < \cdots$, be the series of instants of arrival from the exterior, which constitute a Poisson process of parameter λ.

Let us denote the series of end of service instants and departure instants by $0 < s_1 < s_2 \ldots$ and $0 < d_1 < d_2 < \cdots$ respectively for this queue. With each s_i, we shall associate a random variable X_i such that:

$$X_i = \begin{cases} 0 \text{ if } s_i \text{ coincides with a departure} \\ (d_j, \ j \leqslant i, \text{ exists such that } s_i = d_j), \\ 1 \text{ if } s_i \text{ is a feedback instant.} \end{cases}$$

In our model, the X_i constitute a series of independent random variables identically distributed by the law:

$$q = P\{X_i = 0\}, \quad p = P\{X_i = 1\}.$$

The X_i are equally independent of the series a_i and s_i. Hence

$$P\{s_{i+1} - s_i < x\} = 1 - e^{-\mu x}.$$

As usual let us call $N(t)$ the number of customers in this queue at time t.

The Chapman–Kolmogorov equations of this system are easily written. Let $p(n, t) \equiv P\{N(t) = n\}$; hence:

$$\frac{d}{dt} p(n, t) = \lambda p(n - 1, t) + \mu q p(n + 1, t) - (\lambda + \mu q) p(n, t), \quad n > 0,$$

$$\frac{d}{dt} p(0, t) = \mu q p(1, t) - \lambda p(0, t);$$

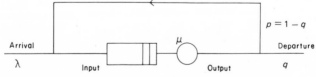

Figure 6.6 M/M/1 queue with feedback

notice that these are identical to those of the ordinary (without feedback) M/M/1 queue with arrival rate λ and service rate μq. We know, therefore, that (Cf. Chapter 1):

(i) $p(n) \equiv \lim\limits_{t \to \infty} p(n,t) = \rho^n(1 - \rho)$, $n \geqslant 0$,   where $\rho = \lambda/\mu q$,

(ii) $\lim\limits_{i \to \infty} P\{N(a_i^-)\} \lim P\{N(d_i^+)\} = p(n)$.

$-$ $F(t)$ is the distribution function of the times between successive *outputs* $(F(t) = P\{s_{i+1} - s_i < t\})$.
Similarly we write:

$$\hat{f}(s) = \int_0^\infty e^{-st} dF(t)$$

and, if it exists, the density:

$$f(t) = \frac{dF(t)}{dt}.$$

$-$ $D(t)$ is the distribution function of the times between successive *departures* $(D(t) = P\{d_{i+1} - d_i < t\})$. The feedback instants $0 < r_i < r_2 < \cdots$, are defined as follows:

$$\{r_1, r_2, \ldots, r_i, \ldots\} = \{s_i | X_i = 1\}$$

and

$$R(t) = P\{r_{i+1} - r_i < t\},$$

$$\hat{r}(s) = \int_0^\infty e^{-st} dR(t), \, r(t) = \frac{dR(t)}{dt}$$

if it exists.

We shall denote the distribution function of the service law by $B(t)$, the density by $b(t)$ and the Laplace transform by $\hat{b}(s)$.

Finally, the series $0 < \alpha_1 < \alpha_2 < \cdots$, of *input* instants is quite simply the ordered set:

$$\{\alpha_1, \alpha_2, \ldots\} = \{a_1, a_2, \ldots\} \cup \{r_1, r_2, \ldots\}.$$

We define:

$$A_i(t) = P\{\alpha_{i+1} - \alpha_i < t\},$$

this definition being particular to a given i, and:

$$A(t) \equiv \lim_{i \to \infty} A_i(t).$$

We shall use:

$$\hat{a}_i(s) = \int_0^\infty e^{-st} dA_i(s), \quad \hat{a}(s) = \int_0^\infty e^{-st} dA(s)$$

to represent the corresponding Laplace transforms.

Input process: calculation of $A(t)$

We are first interested in the properties of effective inputs to the queue. Let E_i be the event whose probability is $1 - A(t)$; then if:

$$r_{ji} = \min\{k \,|\, r_k > \alpha_i\}$$

and:

$$a_{ji} = \min\{k \,|\, a_k > \alpha_i\},$$

one has:

$$E_i = \{(r_{ji} - \alpha_i > t) \quad \text{and} \quad (a_{ji} - \alpha_i > t)\}.$$

Since the arrivals are Poisson, by using the theorem of conditional probabilities one obtains:

$$1 - A_i(t) = e^{-\lambda t}(1 - G_i(t))$$

where

$$G_i(t) = P\{r_{ji} - \alpha_i t \,|\, \alpha_{ji} - \alpha_i > t\}$$

of which the Laplace transform will be denoted by $\hat{g}_i(s)$. From equation (6.4.1), by using the property $A_i(0) = G_i(0) = 0$, one obtains the expression:

$$\hat{a}_i(s) = \frac{\lambda}{\lambda + s}[1 - \hat{g}_i(\lambda + s)] + \hat{g}_i(\lambda + s). \tag{6.4.2}$$

Let us denote the stationary probability of having n customers just after the instant t_0 by $p(n, t_0)$ and let:

$$\Pi_{t_0}(\geq i) = \sum_{n=i}^{\infty} p(n, t_0).$$

The expression for the function $\hat{g}(s)$ is obtained by knowing firstly the state of the system just before t_0 and secondly the number of *departures* occurring during time t. This number depends effectively on the state of the queue at time t, also the only possible events are departures. Therefore we can write:

$$\hat{g}(s) = \sum_{n=0}^{\infty} \Pi_{t_0}(\geq n) p q^n [\hat{b}(s)]^{n+1}$$

if one knows that n departures occur with probability q^n, that there are at least n customers and that the time interval is distributed as the convolution of n service times. But if one is in the stationary state, the distribution of the length of the queue for the input times coincides with that for the departure times (cf. [16]) which is that for 'any' times. Hence $p(n, t)$ is replaced by its value in the stationary state:

$$p(n) = \rho^n(1 - \rho).$$

Hence one obtains:

$$\hat{g}(s) = \frac{p\hat{b}(s)}{1 - q\rho\hat{b}(s)}. \tag{6.4.3}$$

$g(t)$ is easily deduced from this:

$$g(t) = p\mu e^{-(\mu - \lambda)t} \quad \text{and} \quad \hat{g}(\lambda + s) = \frac{p\mu}{\mu + s}$$

which is then substituted into equation (6.4.2) to obtain finally:

$$\hat{a}(s) = \frac{\lambda\mu + s(\lambda + p\mu)}{(\lambda + s)(\mu + s)}$$

or, in a more explicit form:

$$\hat{a}(s) = \frac{p\mu}{\mu - \lambda} \frac{\mu}{\mu + s} + \frac{\mu q - \lambda}{\mu - \lambda} \frac{\lambda}{\lambda + s} \qquad (6.4.4)$$

The inverse transform, with the condition $A(0) = 0$, gives the distribution of the time intervals between successive inputs:

$$A(t) = \frac{p\mu e^{-\mu t}}{u - \lambda} + \frac{(\mu q - \lambda)e^{-\lambda t}}{\mu - \lambda} \qquad (6.4.5)$$

The time intervals separating successive inputs to the queue are therefore distributed following a hyperexponential distribution of order 2 since the coefficients satisfy:

$$\frac{p\mu}{\mu - \lambda} + \frac{\mu q - \lambda}{\mu - \lambda} = 1.$$

The total stream of inputs is not therefore a Poisson stream. This is explained by the interdependence of two possible sources of arrivals. In fact, the time which elapses before feedback depends strongly on the preceding inputs. It is clearly evident that this dependence disappears in the trivial case $p = 0$, and one returns to the case of a Poisson entry process of rate λ from equation (6.4.5).

On the other hand, from equation (6.4.4) which gives $\hat{a}(s)$, the mean value of the times between successive inputs and also its second moment are easily obtained by differentiation. One has:

$$M_1 \equiv \int_0^\infty t\,dA(t) = -\left.\frac{\partial \hat{a}(s)}{\partial s}\right|_{s=0} = \frac{q}{\lambda},$$

$$M_2 \equiv \int_0^\infty t\,dA(t) = \left.\frac{\partial^2 \hat{a}(s)}{\partial s^2}\right|_{s=0} = \frac{2q}{\lambda^2} - \frac{2p}{\lambda\mu}.$$

It will be useful, in the following, to establish a relation between these two first moments and certain values of the function $\hat{g}(s)$. From equation (6.4.3), it is easy to show that:

$$M_1 = \frac{1}{\lambda}[1 - \hat{g}(\lambda)], \qquad (6.4.6)$$

$$M_2 = \frac{2}{\lambda^2}[1 - \hat{g}(\lambda)] + \frac{2}{\lambda}\left.\frac{\partial \hat{g}(\lambda + s)}{\partial s}\right|_{s=0} \qquad (6.4.7)$$

If the square of the variation coefficient of the times between successive inputs is denoted by Ka, one obtains:

$$Ka = \frac{M_2 - M_1^2}{M_1^2} = \frac{1 - \hat{g}^2(\lambda) + 2\lambda \left.\frac{\partial \hat{g}(\lambda + s)}{\partial s}\right|_{s=0}}{[1 - \hat{g}(\lambda)]^2}.$$

6.4.2. Output process

We wish to determine the distribution law of the time intervals between successive ends of service whose Laplace transform is denoted by $\hat{f}(s)$.

Let us denote the series of end of service instants by $(s_n)_{n \in N}$ and introduce the following variable:

$X_n = 1$ if, at the instant s_n, the customer who has just finished his service is reintroduced to the queue.

$X_n = 0$ if, at the instant s_n, this customer definitely leaves the system.

Evidently, one has:

$$P\{X_n = 1\} = p,$$
$$P\{X_n = 0\} = q.$$

From the property of conditional probabilities, one can write:

$$P\{s_{n+1} - s_n \leqslant t\} = pP\{s_{n+1} - s_n \leqslant t | X_n = 1\} \\ + qP\{s_{n+1} - s_n \leqslant t | X_n = 0\}. \tag{6.4.8}$$

At instants $(s_n + \varepsilon)$, $n \to \infty$, the state of the queue is given by the stationary probability of the system (these are the instants following an output).

Following the nth service, if the customer leaves the system, the queue is empty with probability $\pi(0) = 1 - \rho$ (this probability is that of the stationary state (cf. Chapter 1)).

The duration of the interval $s_{n+1} - s_n$ is distributed as the convolution of the two following functions:

– the distribution of the times which precede the first exterior arrival and
– the distribution of the $(n + 1)$th service.

But if, at the instant $(s_n + \varepsilon)$, the customer is reintroduced into the queue, the interval $s_{n+1} - s_n$ is simply the time necessary for the execution of the $(n + 1)$th service, independently of the state of the system, since this contains at least the reintroduced customer.

Let us express (6.4.8) with the help of Laplace transforms following the previous comments. One has:

$$\hat{f}(s) = p\hat{b}(s) + \left[qp(0)\frac{\lambda}{\lambda + s}\hat{b}(s) + (1 - p(0))\hat{b}(s) \right], \tag{6.4.9}$$

$$\hat{f}(s) = \hat{b}(s)[1 - qp(0)] + qp(0)\hat{b}(s)\frac{\lambda}{\lambda + s},$$

which gives:

$$\hat{f}(s) = \frac{p\mu}{\mu - \lambda} \frac{\mu}{\mu + s} + \frac{\mu q - \lambda}{\mu - \lambda} \frac{\lambda}{\lambda + s}. \tag{6.4.10}$$

Form this last expression, one concludes that the end of service instants do not form a Poisson process except in the case of a zero probability of feedback. Furthermore, we have shown the equality of the functions $\hat{a}(s)$ and $\hat{f}(s)$. The process defined by the *input* instants is that defined by the *output* instants which have exactly the same distribution.

Comment

This identity of distributions was established by R. Disney and D. MacNickle [16] who obtained $\hat{f}(s)$ given by relation (5.3.9) by demonstrating the following result:

$$\tilde{p}(0) = q\left(1 - \frac{\lambda}{\mu}\right) = q p(0),$$

$$\tilde{p}(j) = \left(1 - \frac{\lambda}{\mu q}\right)\left(\frac{\lambda}{\mu q}\right)^{j-1}\left(p + \frac{\lambda}{\mu}\right), \quad j = 1, 2, \ldots,$$

where they designate the probability of being in state j immediately after the end of service by $\tilde{p}(j)$. Now this seems to contradict the state probability which we have used. This relative difference arises from the method used to calculate the number of customers. If this number taken at any instant t is denoted by $N(t)$, we have

$$p(j) = P\{N(s_n - \varepsilon) - 1 = j\}$$

and

$$\tilde{p}(j) = P\{N(s_n + \varepsilon) = j\}.$$

The probability p, therefore, does not measure the customer who finishes his service at the instant s_n. The probability $\tilde{p}$ takes account of this customer in the case of a return to the queue.

6.4.3. Departure process

Denote the series of departure instants by $(d_j)_{j \in \mathbb{N}}$. The corresponding distribution function is

$$D(t) = P\{d_{n+1} - d_n \leqslant t\},$$

and its density function is $d(t)$.

The density of the conditional distribution $P\{s_{n+1} - s_n \leqslant t | X_n = 0\}$ introduced previously is designated by $\alpha(t)$ and the Laplace transform of $\alpha(t)$ by $\hat{\alpha}(s)$. The distribution between two ends of service, knowing that the first is a feedback,

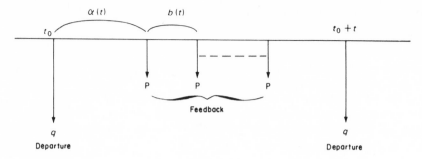

Figure 6.7

is denoted by

$$P\{s_{n+1} - s_n \leqslant t | X_n = 1\}$$

and has a similar density function $b(t)$.

Knowing $\alpha(t)$ and $b(t)$ and the number of returns to the queue occurring during an interval of length t, we can express the required density function. One has the schematic representation of Figure 6.7.

If the two departures are consecutive (without feedback), from the same definition of $\alpha(t)$ we have:

$$d(t) = q\alpha(t).$$

And for a positive number n of feedbacks:

$$d(t) = qp^n\alpha(t)*b^{*n}(t),$$

where $*$ designates a convolution operation.

By summing over the possible values of n and using Laplace transforms, one obtains:

$$\hat{d}(s) = \sum_{n=0}^{\infty} qp^n\hat{\alpha}(s)\hat{b}^n(s) = \frac{q\hat{\alpha}(s)}{1 - p\hat{b}(s)},$$

with

$$\hat{\alpha}(s) = p(0)\frac{\lambda}{\lambda + s}\,\hat{b}(s) + (1 - p(0))\hat{b}(s).$$

By substituting the expressions for $\hat{b}(s)$ and $p(0)$, this becomes:

$$\hat{d}(s) = \frac{\lambda}{\lambda + s}. \tag{6.4.11}$$

And hence:

$$D(t) = 1 - e^{-\lambda t}.$$

The times between successive departures from the system are, therefore, random variables distributed with an exponential law of parameter λ.

6.4.4 Feedback process

This process has been studied for a long time without obtaining results on account of its complexity. It is clearly evident that the distribution of times between feedbacks will be easily obtained if the end of service instants form a renewal process. In fact, one could use the reasoning of Section 6.4.1 to deduce it in an analogous manner:

$$\hat{r}(s) = \frac{p\hat{b}(s)}{1 - q\hat{a}(s)}$$

(recall that $\hat{r}(s) = \int_0^\infty e^{-st} dR(t)$, if $R(t)$ is the distribution function of the times between sucessive feedbacks). Unfortunately, this is not the case for, as we shall see later, the series of intervals $[s_n, s_{n+1}]$ is not a series of independent variables which explains the difficulty.

However, we present a relatively simple method which uses the Markov property to allow an explicit form of the Laplace function $\hat{r}(s)$ to be obtained.

From the definition of equilibrium probability given in Section 6.4.1, the queue is in state $i(i = 1, 2, \dots)$ at instant $t_0 + \varepsilon$, with a probability $p(i-1) = \rho^{i-1}(1-\rho)$, if t_0 is the instant of any feedback.

Knowing the possible states of the system, the function $\hat{r}(s)$ can be put in the form:

$$\hat{r}(s) = \sum_{i=1}^{\infty} p(i-1)\alpha(i) \qquad (6.4.12)$$

in which $\alpha(i)$ is the Laplace transform of the time which separates two feedbacks, *knowing that* the system is in state i at the beginning of the interval.

It is this function $\alpha(i)$ which we wish to study. First recall the following result: 'the superposition of $k(k \in N)$ independent Poisson streams of respective parameters $a_1, a_2, \dots, a_k$ is again a Poisson stream of which the parameter a is the sum of the parameters $a_i(i = 1, \dots, k)$'.

Since the initial instant, t_0, is chosen as a feedback instant, the queue contains at least one customer. The first event which occurs after t_0, say at $t_0 + T$, can be of two kinds: an exterior arrival or an end of service. The length of the interval between the initial instant and the instant of occurrence of this event is distributed as the combination of two exponential distributions of which the parameters are λ and μ. From the preceding recollection, the Laplace transform of its density function is therefore: $(\lambda + \mu)/(\lambda + \mu + s)$.

The event which occurs at the instant $t_0 + T$ is an arrival from the exterior with a probability $\lambda/(\lambda + \mu)$. The queue is in state $i + 1$ if it contains i customers at the instant t_0. But with a probability $\mu q/(\lambda + \mu)$ there is a departure and in this case the state of the queue is $(i - 1)$.

Finally with a probability $\mu p/(\lambda + \mu)$ this event is a feedback and it is not useful to consider the state of the queue.

This can be summarized by the following equations:

$$\alpha(i) = \frac{\lambda + \mu}{\lambda + \mu + s}\left[\frac{\lambda}{\lambda + \mu}\alpha(i+1) + \frac{\mu q}{\lambda + \mu}\alpha(i-1) + \frac{\mu p}{\lambda + \mu}\right], \quad i > 0.$$

We also need to express $\alpha(0)$ and evidently one has:

$$\alpha(0) = \frac{\lambda}{\lambda + s}\alpha(1)$$

since if the queue is empty at the initial instant, the only possible subsequent event is an arrival from the exterior.

After an evident simplification, these equations can be put in the form:

$$(\lambda + \mu + s)\alpha(i) = \lambda\alpha(i+1) + \mu q\alpha(i-1) + \mu p, \quad i > 0 \qquad (6.4.13)$$

$$(\lambda + s)\alpha(0) = \lambda\alpha(1). \qquad (6.4.14)$$

Then (6.4.13) and (6.4.14) are multiplied by the probabilities $p(i-1)$ and $p(0)$ respectively. Then all the equations obtained are summed from $i = 1$ to obtain the function $\hat{r}(s)$. By using the relation $p(i) = \rho p(i-1) = (\lambda/\mu q)p(i-1)$, after some simplifications one obtains:

$$(\mu p + s)\hat{r}(s) = \mu p - \frac{\mu q}{\lambda}sp(0)\alpha(0). \qquad (6.4.15)$$

It now remains to determine the term $\alpha(0)$.

For that we return to equation (6.4.13) in which the following change of variable is made:

$$\alpha(i) = \beta(i) + \frac{\mu p}{\mu p + s}.$$

This allows us to obtain new equations in $\beta(i)$:

$$(\lambda + \mu + s)\beta(i) = \lambda\beta(i+1) + \mu q\beta(i-1), \quad i > 0,$$

which has a known solution. It is of the form:

$$\beta(i) = ax_1^i + bx_2^i, \quad i \geqslant 0,$$

where a and b are coefficients to be determined and x_1 and x_2 are the roots of the second order equation:

$$\lambda x^2 - (\lambda + \mu + s)x + \mu q = 0.$$

One has therefore:

$$x_1 + x_2 = \frac{\lambda + \mu + s}{\lambda},$$

$$x_1 x_2 = \frac{\mu q}{\lambda} = \frac{1}{q} > 1.$$

The product of the roots is greater than unity if the ergodicity condition of the system is satisfied. From this one deduces that one of the two roots, for example x_2, is also greater than unity. But, for an infinite number of customers, it is evident from equation (6.4.13) that we must have:

$$\lim_{i \to \infty} \alpha(i) = \frac{\mu p}{\mu p + s}.$$

Hence:

$$\lim_{i \to \infty} \beta(i) = 0.$$

In order to observe this convergence it is necessary that coefficient b of the greater of the roots is zero. It follows that:

$$\beta(i) = ax_1^i, \quad i \geqslant 0, \tag{6.4.16}$$

with

$$x_1 = \frac{\lambda + \mu + s - \sqrt{(\lambda + \mu s)^2 - 4\lambda\mu q}}{2\lambda}.$$

From which the following system is deduced:

$$\alpha(0) = a + \frac{\mu p}{\mu p + s},$$

$$\alpha(1) = \frac{\lambda + s}{\lambda}\alpha(0) = ax_1 = \frac{\mu p}{\mu p + s},$$

which allows us to determine the coefficient a and the function $\alpha(0)$. One obtains

$$\alpha(0) = \frac{\mu p}{\mu p + s}\frac{1 - x_1}{1 - x_1 + s\lambda}, \tag{6.4.17}$$

$$a = \frac{\mu p}{\mu p + s}\frac{-s/\lambda}{1 - x_1 + s/\lambda}.$$

The density function of the times between successive feedbacks is, therefore, entirely determined by its Laplace transform which is written:

$$\hat{f}(s) = \frac{\mu p}{\mu p + s}\left[1 - \mu qs\pi(0)\frac{(1 - x_1)}{(\mu p + s)\lambda(1 - x_1) + s}\right]. \tag{6.4.18}$$

From equation (6.4.18) it is also possible to verify that the feedback rate is $\lambda p/q$. In effect one has:

$$\left.\frac{\partial \hat{f}(s)}{\partial s}\right|_{s=0} = -\frac{1}{\mu p} - \frac{\mu qp(0)}{\lambda\mu p} = -\frac{q}{\lambda p}.$$

The input rate is the sum of the two arrival rates since:

$$\frac{\lambda}{q} = \frac{\lambda p}{q} + \lambda.$$

Equation (6.4.18) proves equally that the *feedback process* is not a Poisson process. In fact, if the function $\hat{r}(s)$ can be written as the Laplace transform of an exponential density, one then has the identity:

$$\hat{r}(s) = \frac{\lambda p/q}{\lambda p/q + s} = \frac{\mu p}{\mu p + s}\left[1 - \mu q s p(0)\frac{(1 - x_1)}{(\mu p + s)[\lambda(1 - x_1) + s]}\right]$$

which is impossible given the purely irrational form of x_1.

In summary, we have determined all the distributions of the different processes which characterize the model of a M/M/1 queue with feedback. Up to now, we have shown that the only Poisson streams in this system are the arrival and departure streams.

We shall be able to show an additional property; the input, output and feedback streams are not even renewal processes.

6.4.5. Flows in Jackson networks

Consider R: $i \leftrightarrow j$ the equivalence relation between queues defined by: 'i and j belong to the same equivalence class if it is possible to go to queue j from queue i and it is possible to go to queue i from queue j'. The equivalence classes which we shall call irreducible sub-networks are formed at stations which cannot be attained once the customers have left. One can show that only flows which connect irreducible sub-networks are Poisson processes. All the interior flows are neither Poisson nor renewal. Another very important property can be demonstrated; the probability distributions of the time intervals between inputs and between end of service instants are identical.

These last results are again true in more general networks having a product form solution. Proofs of these theorems can be found in the articles cited in the bibliography.

EXERCISES

1. Show by the reversibility technique that the output process of an M/M/1/N queue (of which the capacity is limited to N) is Poisson if the customers who are refused are sent directly to the output. Then show that if the refused customers are not counted in the output process this is not Poisson.

2. Show that in a Jackson network in which a customer leaving a station has zero probability of returning, all the streams are Poisson and the network as a whole takes the form of a reversible process.

3. Consider a queue for which the service times are independent and which have service laws taking one of the four forms imposed by the BCMP theorem (see Chapter 4). Show that the queue is reversible if the arrival process is Poisson. Hence deduce that the output process is Poisson.

4. Show that the result giving the necessary and sufficient condition of

reversibility for a stationary Markov process is still true for a stationary Markov chain. The probability $q(k,k')$ of passing from state k to state k' replaces the transition rate.

COMMENT: For the Markov chain embedded in the Markov process $N(t)$ we have:

$$q(k,k') = \frac{d(k,k')}{\sum\limits_{k'} d(k,k')}.$$

5. Show that a stationary Markov chain is reversible if and only if the transition probabilities satisfy:

$$q(k_1,k_2)q(k_2,k_3)\cdots q(k_n,k_1) = q(k_1,k_n)q(k_n,k_{n-1})\cdots q(k_2,k_1),$$

for all finite series of states $k_1,\ldots,k_n$.

Show the same property for stationary processes by replacing $q(k_i,k_j)$ by $d(k_i,k_j)$.

6. Consider an M/M/1 queue. Now consider the corresponding process at unoccupied places in the queue. When a customer enters the queue, an unoccupied space leaves and conversely when a customer leaves the queue an unoccupied space enters it. Show that this new queue is not quasi-reversible.

7. Consider two M/M/1 queues in series and the unique equivalent queue (identical output processes and the number of customers in the two queues equal to the number of customers in the equivalent queue). Show that this new queue is not quasi-reversible although the input and output processes may be independent and Poisson.

BIBLIOGRAPHY

In Chapter 6 we study some properties of streams in queueing networks and more particularly when the arrival processes from the exterior are Poisson and the service times exponentially distributed (Jackson networks [1] and [2]).

The output process of a queue has been studied from 1956 by Burke [3] then by Reich [4] and Finch [5]. There have been numerous subsequent works on this subject; several are cited in [6] [7] [8] and [9]. These studies have led to an examination of product form networks; these have led to the BCMP theorem [12] studied in Chapter 4, above all following Chandy's studies of local balance equations [10] extended in [11].

Section 6.3 is devoted to the study of reversible and quasi-reversible networks introduced by Kelly [13].

Streams in the M/M/1 system with feedback have posed many questions which are now resolved [14] to [18]. In the case of a general network, numerous properties became known [19] to [38] for networks whose solution is in the form of a product (product of the marginal probabilities).

1. Jackson, R. (1963). Jobshop-like queueing systems, *Management Science*, **10**, 131–142.
2. Gordon, W. I., and Newell, G. F. (1967). Closed queueing systems with exponential servers, *Oper. Res.*, **15**, 154–165.

3. Burke, P. J. (1956). The output of a queueing system, *Oper. Res.*, **4**, 699–704.
4. Reich, E. (1957). Waiting times when queues are in tandem, *Ann. Math. Statist.*, **28**, 768–773.
5. Finch, P. O. (1959). The output process of the queueing system M/G/1, *J. Roy. Stat. Soc., Ser. B.*, **21**, 375–380.
6. Disney, R. L., Cherry, W. P. (1974). *Some Topics in Queueing Network Theory, in Mathematical Methods for Queueing Theory*, 23–44, Springer Verlag.
7. Disney, R. L., Farrel, R. L., and De Morais P. R. (1973). A characterization of M/G/1/N queues with renewal departure processes, *Management Sciences*, **19**, 1222–1228.
8. Daley, D. J. (1968). The correlation structure of the output process of some single server queueing systems, *Ann. Math. Statist.*, **39**, 1007–1019.
9. Laslett, G. M. (1975). Characterizing the finite capacity GI/M/1 queue with renewal output, *Management Sciences*, **22**, 106–110.
10. Chandy, K. M. (1972). The analysis and solutions for general queueing networks, *Proc. Sixth Annual Princeton Conference on Information Sciences and Systems, Princeton University*.
11. Chandy, K. M., Howard, J. H., and Towsley, D. F. (1977). Product form and local balance in queueing networks, *J. ACM*, **24**, 250–263.
12. Basket, F., Chandy, K. M., Muntz, R. R., Palacios, F. G. (1975). Open closed and mixed networks of queues with different classes of customers, *J. ACM*, **22**, 248–260.
13. Kelly, F. P. (1979). *Reversibility and Stochastic Networks*, Wiley.
14. Takacs, L. (1963). A single server queue with feedback, Bell systems, *Technical Journal*, **42**, 505–519.
15. Burke, P. J. (1976). Proof of a conjecture on the interarrival time distribution in an M/M/1 queue with feedback, *IEEE Trans. on Communications*, **24**, 575–576.
16. Disney, R. L., and Mc Nickle, D. C. (1977). *The M/G/1 queue with instantaneous Bernoulli feedback* – Research report 77–3, University of Michigan.
17. Foley, R. D. (1977). *Queues with feedback,* – Ph. D. Dissertation, University of Michigan.
18. Pujolle, G. (1980). Réseaux de files d'attente à forme produit, *RAIRO verte*, **14**, 317–330.
19. Kelley, F. P. (1976). Networks of queues, *Adv. appl. prob.*, **8**, 416–432.
20. Beutler, F. J., and Melamed, B. (1978). Decomposition and customer streams of feedback networks of queues in equilibrium, *Oper. Res.*, **26**, 1059–1072.
21. Pujolle, G., and Soula, C. (1979). A study of flows in queueing networks and an approximate method for solution, *Proc. 4th International Congress on Modelling*, Vienna.
22. Labetoulle, J., Pujolle, G., and Soula, C. (1981). Distribution of the flows in a general Jackson network, *Math. Oper. Res.*, **6**, 173–185.
23. Barbour, A. D. (1976). Networks of queues and the method of stages, *Adv. Appl. Prob.*, **8**, 584–591.
24. Stoyan, D. (1978). Queueing networks – Insensitivity and a heuristic approximation, – Elektion, *Informations Verarbeit, Kybernetik*, **14**.
25. Schassberger, R. (1978). The insensitivity of stationary probabilities in networks of queues, *Adv. Appl. Prob.*, **10**, 906–912.
26. Schassberger, R. (1978). Insensitivity of steady-state distributions of generalized semi-Markov processes with speeds, *Adv. Appl. Prop.*, **10**, 836–851.
27. Bremaud, P. (1981). *Dynamical Point Processes and Ito Systems in Communications and Queueing*, Springer Verlag.
28. Bremaud, P. (1979). *A counter example to a converse of Burke–Reich theorem*, Research report ENSTA.
29. Muntz, R. R. (1973). Poisson departure processes and queueing networks, *Proc. 7th Annual Conf. Information Sciences and Systems, Princeton University*, 435, 440.

30. Gelenbe, E., and Muntz, R. R. (1976). Probabilistic models of computer systems, *Acta Informatica*, **7**, 35–60.
31. Melamed, B. (1979). On Poisson traffic processes in discrete-state Markovian systems with applications to queueing theory, *Adv. Appl. Prob.*, **11**, 218–239.
32. Sevcik, K. and Mitrani, I. (1979). The distribution of queueing network states at input and output instants, *Proc. Int. Symp. Performance Computer Systems*, Vienna.
33. Lavenberg, S., and Reiser, M. (1979). *The state seen by an arriving customer in closed multiple chain queueing networks*, – Research report IBM Yorktown Heights.
34. Walrand, J., and Varatya, P. (1978). *The outputs of Jacksonian networks are Poissonian*, – Research report ERL-M 78/60, – University of Berkeley.
35. Kelly, F. P. (1975). Networks of queues with customers of different types, *J. Appl. Prob.*, **12**, 542–554.
36. Pellaumail J. (1978). Régimes stationnaires quand les routages dépendent de l'état, *Actes du 1er colloque AFCET-SMF de mathématiques appliquées*, Palaiseau.
37. Le Ny, L. M. (1979). *Étude analytique de réseaux de files d'attente multi-classes à routages variables*, Thèse de 3è cycle, Rennes I.
38. Lam, S. S. (1976). Store-and-forward buffer requirements in a packet switching network, *IEEE Trans. Com.*, **24**, 394, 399.

Appendix

Principal formulae used in models of queueing networks.

M/M/1 queue

$$p(n) = \rho^n(1 - \rho) \qquad \rho = \lambda/\mu$$
$$E(N) = \rho/(1 - \rho)$$
$$\sigma_N^2 = \rho/(1 - \rho)^2$$
$$E(L) = \rho^2/(1 - \rho)$$
$$E(w) = \rho/\mu(1 - \rho)$$
$$E(T) = 1/\mu(1 - \rho)$$

distribution of response time $T(x) = \mu(1 - \rho)e^{-\mu(1 - \rho)x}$

M/M/1/K queue

$$p(n) = \frac{(1 - \rho)\rho^n}{1 - \rho^{k+1}} \quad \text{if } \lambda \neq \mu$$

$$p(n) = \frac{1}{K + 1} \quad \text{if } \lambda = \mu$$

Probability that a customer is lost $p(K) = (1 - \rho)\rho^K/(1 - \rho^{K+1})$.

$$E(N) = \rho\left[\frac{1 - (K + 1)\rho^K + K\rho^{K+1}}{(1 - \rho)(1 - \rho^{K+1})}\right] \quad \text{if } \lambda \neq \mu$$

$$E(N) = \frac{K}{2} \qquad \text{if } \lambda = \mu$$

$$E(L) = E(N) - (1 - p(0))$$
$$E(W) = E(L)/\lambda(1 - p(K))$$
$$E(T) = E(N)/\lambda(1 - \rho(K))$$

168

M/M/C queue

$$p(n) = p(0)\frac{(C_\rho)^n}{n!} \qquad\qquad \text{if } n \leqslant C, \rho = \frac{\lambda}{u}$$

$$p(n) = p(0)\frac{\rho^n}{C^{n-c}\cdot C!} \qquad\qquad \text{if } n \geqslant C$$

$$p(0) = \left[\sum_{k=1}^{C-1}\frac{(C\rho)^k}{k!} + \frac{(\rho)^C}{c!}\frac{1}{1-\rho/c}\right]^{-1}$$

Probability that all the servers are occupied

= Erlang's formula C

$$= \mathscr{E}(C,\rho) = \frac{\rho^C}{C!}\Bigg/\left[\frac{\rho^C}{C!} + (1-\rho)\sum_{k=0}^{C-1}\frac{\rho^k}{k!}\right]$$

$$E(N) = \rho + E(L)$$

$$E(L) = \frac{\rho^C\lambda\mu}{(C-1)!(C\mu-\lambda)^2}p(0)$$

$$E(W) = E(L)/\lambda$$

$$E(T) = E(N)/\lambda$$

M/M/C/C and M/GI/C/C queues

$$p(n) = \frac{\rho^n/n!}{1 + \rho + \rho^2/2! + \cdots + \rho^C/C!}$$

Probability that a customer is rejected = Erlang's formula

$$= p(C) = \frac{\rho^C/C!}{1 + \rho + \rho^2/2! + \cdots + \rho^C/C!}$$

$$E(N) = \rho(1 - \rho(C))$$

$$E(L) = 0$$

$$E(W) = 0$$

$$E(T) = \frac{1}{\mu}$$

M/M/C/K queue

$$p(n) = p(0)\frac{\rho^n}{n!} \qquad\qquad \text{for } n \leqslant C$$

$$p(n) = p(0)\frac{\rho^C}{C!}\left(\frac{\rho}{C}\right)^{n-C} \qquad \text{for } C \leqslant n \leqslant K$$

$$p(0) = \left[\sum_{n=0}^{C}\frac{\rho^n}{n!} + \frac{\rho^C}{C!}\sum_{n=1}^{K-C}\left(\frac{\rho}{C}\right)^n\right]^{-1}$$

$$E(N) = E(L) + \sum_{n=0}^{C-1} np(n) + C\left(1 - \sum_{n=0}^{C-1} p(n)\right)$$

$$E(L) = p(0)\frac{\rho^{C+1}/C}{C!(1 - \rho/C)^2}\left[1 - \left(\frac{\rho}{C}\right)^{K-C+1} - K - C + 1\left(\frac{\rho}{C}\right)^{K-C}\left(1 - \frac{\rho}{C}\right)\right]$$

$$E(W) = E(L)/\lambda(1 - p(K))$$

$$E(T) = E(N)/\lambda(1 - p(K))$$

M/M/∞ and M/GI/∞ queues

$$p(n) = e^{-\rho}\frac{\rho^n}{n!}$$

$$E(N) = \rho$$

$$E(L) = 0$$

$$E(W) = 0$$

$$E(T) = \frac{1}{\mu}$$

M/D/1 queue

$$E(N) = \frac{2\rho - \rho^2}{2(1 - \rho)}$$

$$E(L) = \frac{\rho^2}{2(1 - \rho)}$$

$$E(W) = \frac{\rho}{2\mu(1 - \rho)}$$

$$E(T) = \frac{2 - \rho}{2\mu(1 - \rho)}$$

Mean number left in the queue by a departing customer $= (\rho/(1 - \rho)) - \rho^2/2(1 - \rho)$

M/E_k/1 queue

$$E_k = \text{Erlang-}k \text{ distribution}$$

$$p(n) = (1 - \rho)\sum_{i=1}^{k}\prod_{\substack{j=1 \\ j\neq i}}^{k}\frac{1}{1 - x_i/x_j}$$

where x_i, $i = 1,\ldots,k$ are the roots of the equation $k\mu - \lambda(x + x^2 + \cdots + x^k) = 0$

$$E(N) = \rho\left[1 + \frac{\rho(1 + 1/k)}{2(1 - \rho)}\right]$$

$$E(L) = \frac{\rho^2}{2(1-\rho)}(1 + 1/k)$$

$$E(W) = \frac{\rho}{2\mu(1-\rho)}(1 + 1/k)$$

$$E(T) = \frac{1}{\mu}\left[1 + \frac{\rho(1 + 1/k)}{2(1-\rho)}\right]$$

M/GI/1 queue

$$p(j) = \Pi_j = \Pi_j^*$$

$$E(N) = \rho\left[1 + \frac{\rho(1 + Ks)}{2(1-\rho)}\right]$$

$$E(L) = \frac{\rho^2(1 + Ks)}{2(1-\rho)}$$

$$E(W) = \frac{\rho(1 + Ks)}{2\mu(1-\rho)}$$

$$E(T) = \frac{1}{\mu}\left[1 + \frac{\rho(1 + Ks)}{2(1-\rho)}\right]$$

$$P(x) = \sum p(x)x^n = \frac{(1-\rho)(1-x)\mathscr{L}f(\lambda(1-x))}{\mathscr{L}f(\lambda(1-x)) - x}$$

where f is the service time distribution.

GI/M/1 queue (particular case E_k/M/1)

$$p(0) = 1 - \rho$$
$$p(n) = \rho(1 - \sigma)\sigma^{k-1}$$

where σ is the smallest root of the polynomial $f(\mu(1 - x))$
(all the roots satisfy $|x_i| < 1$ except $|x_0| > 1$)

$$E(N) = \frac{\rho}{1 - \sigma}$$

$$E(L) = \frac{\sigma\rho}{1 - \sigma}$$

$$E(W) = \frac{\sigma}{\mu(1 - \sigma)}$$

$$E(T) = \frac{1}{\mu(1 - \sigma)}$$

GI/GI/1

We give the principal approximations which have been found to date together some bounds.

$$E(N) = \rho\left[1 + \frac{\rho(Ka + Ks)}{2(1 - e)}\right] \qquad \text{Kingman Diffusion}$$

$$E(N) = \frac{\rho}{1 - \hat{\rho}} \quad \text{where} \quad \hat{\rho} = \exp\frac{2(1 - e)}{\rho Ka + Ks} \qquad \text{Diffusion}$$

$$E(N) = \rho\left[1 + \frac{\rho Ka + Ks}{2(1 - e)}\right] \qquad \text{Diffusion}$$

$$E(N) = \rho + Ka\,KsE(N)_{M/M/1} + Ka(1 - Ks)_{M/D/1}$$
$$\qquad\qquad + Ks(1 - Ka)E(N)_{D/M/1} \qquad \text{Page}$$

$$E(N) \leqslant \rho\left[1 + \frac{Ka/d^2 + Ks/\mu^2}{2(1 - e)}\right] \qquad \text{Kingman}$$

$$E(N) \geqslant \rho\left[1 + \frac{Ka/d^2 + Ks/\mu^2}{2(1 - e)} - \frac{1}{2}(Ka + \rho)\right]$$

$$E(N) \geqslant \rho\left[1 + \rho\frac{\rho Ks + \rho - 2}{2(1 - e)}\right] \qquad \text{Marchal}$$

$$E(N) = \rho\left[1 + \frac{\rho(Ka + Ks)}{2(1 - e)}g\right]$$

where

$$g = \begin{cases} \exp\left[\dfrac{-2(1 - e)(1 - Ka)^2}{3\rho(Ka + Ks)}\right] & \text{if } Ka < 1 \\[3mm] \exp\left[\dfrac{-(1 - \rho)(Ka - 1)}{Ka + 4Ks}\right] & \text{if } Ka \geqslant 1 \end{cases}$$

GI/GI/C

Principal approximations

$$E(N) = C\rho + \frac{Ka + \rho^2 Ks}{2(1 - e)} \qquad \text{Kingman}$$

$$E(N) = C\rho + \frac{(Ka + Ks)\rho\sqrt{2(C + 1)}}{2(1 - e)} \qquad \text{Sakasegawa}$$

$$E(N) = \rho\left[1 + \frac{Ka/\gamma^2 + Ks/C^2\mu^2}{2(1 - e)}\right] \qquad \text{if } \rho \text{ is close to 1}$$

Index